INNOVATIVE TECHNOLOGY SERIES
INFORMATION SYSTEMS AND NETWORKS

D0101334

Mobile Agents for

Telecommunication
Applications

edited by
Eric Horlait

HPS
HERMES PENTON SCIENCE

First published in 2000 by Hermes Science Publications, Paris
First published in 2002 by Hermes Penton Ltd
Derived from Networking and Information Systems Journal, *Mobile Agents for Telecommunication Applications*, Vol. 3, No. 5–6.

Hermes Penton Science
120 Pentonville Road
London N1 9JN

© Hermes Science Publications, 2000
© Hermes Penton Ltd, 2002

The right of Eric Horlait to be identified as the editor of this work has been asserted by him in accordance with the Copyright, Designs and Patents Act 1988.

British Library Cataloguing in Publication Data

A CIP record for this book is available from the British Library.

ISBN 1 9039 9628 7

Typeset by Saxon Graphics Ltd, Derby
Printed and bound in Great Britain by Biddles Ltd, Guildford and King's Lynn
www.biddles.co.uk

Contents

Foreword

This publication is concerned with mobile agents for telecommunication applications. Papers have been selected from those presented during MATA'00 in Paris, the MATA reviewers having reviewed this second version of the papers presented here.

Mobile agents refer to self-contained and identifiable computer programs that can move within the network and can act on behalf of the user or another entity. Most current research work on the mobile agent paradigm has two general goals: reduction of network traffic and asynchronous interaction. These two goals stem directly from the desire to reduce information overload and to efficiently use network resources.

There are certainly many motivations for the use of a mobile agent paradigm. However, intelligent information retrieval, network and mobility management, and network services are currently the three most cited application targets for a mobile agent system.

We provide in this publication an overview of how mobile codes could be used in networking. A huge field of application is now open and a research community really exists. We have tried here to illustrate this emerging application domain of mobile agents and mobile code.

Eric Horlait

Chapter 1

Implementing secure distributed computing with mobile agents

Gregory Neven, Erik Van Hoeymissen,
Bart De Decker and Frank Piessens
Department of Computer Science, KU Leuven, Belgium

1. Introduction

Secure distributed computing (SDC) addresses the problem of distributed computing where some of the algorithms and data that are used in the computation must remain private. Usually, the problem is stated as follows, emphasizing privacy of data. Let f be a publicly known function taking n inputs, and suppose there are n parties (named p_i, $i = 1 \dots n$), each holding one private input x_i. The n parties want to compute the value $f(x_i, \dots, x_n)$ without leaking any information about their private inputs (except of course the information about x_i that is implicitly present in the function result) to the other parties. An example is voting: the function f is addition, and the private inputs represent yes ($x_i = 1$) or no ($x_i = 0$) votes. In case you want to keep an algorithm private, instead of just data, you can make f an interpreter for some (simple) programming language, and you let one of the x_i be an encoding of a program.

In descriptions of solutions to the secure distributed computing problem, the function f is usually encoded as a boolean circuit, and therefore secure distributed computing is also often referred to as *secure circuit evaluation*.

It is easy to see that an efficient solution to the secure distributed computing problem would be an enabling technology for a large number of interesting distributed applications across the Internet. Some example applications are: auctions ([NIS 99]), charging for the use of algorithms on the basis of a usage count ([SAN98, SAN 98b]), various kinds of weighted voting, protecting mobile code integrity and privacy ([SAN 98, LOU 99]).

Secure distributed computing is trivial in the presence of a globally trusted third party (TTP): all participants send their data and code to the TTP (over a secure channel), the TTP performs the computation and broadcasts the results. The main drawback of this approach is the large amount of trust needed in the TTP.

However, solutions without a TTP are also possible. Over the past two decades, a fairly large variety of solutions to the problem has been proposed. An overview is given by Franklin [FRA 93] and more recently by Cramer [CRA 99]. These solutions differ from each other in the cryptographic primitives that are used, and in the class of computations that can be performed (some of the solutions only allow for specific kinds of functions to be computed). The main drawback of these solutions is the heavy communication overhead that they incur. For a case study investigating the communication overhead in a concrete example application, we refer the reader to [NEV 00].

Mobile agents employing these cryptographic techniques can provide for a trade-off between communication overhead and trust. The communication overhead is alleviated if the communicating parties are brought close enough together. In our approach, every participant sends its representative agent to a trusted execution site. The agent contains a copy of the private data x_i and is capable of running a SDC-protocol. Different participants may send their agents to different sites, as long as these sites are located closely to each other. Of course, a mobile agent needs to trust his execution platform, but we show that the trust requirements in this case are much lower than for a classical TTP. Also, in contrast with protocols that use unconditionally TTPs, the trusted site is not involved directly. It simply offers a secure execution platform: i.e. it executes the mobile code correctly, does not spy on it and does not leak information to other mobile agents. Moreover, the trusted host does not have to know the protocol used between the agents. In other words, the combination of mobile agent technology and secure distributed computing protocols makes it possible to use a *generic* TTP that, by offering a secure execution platform, can act as TTP for a wide variety of protocols in a uniform way. A detailed discussion of the use of mobile agent technology for advanced cryptographic protocols is given in Section 3.

The combination of cryptographic techniques for secure computing and mobile code has been investigated from another point of view by Sander and Tschudin ([SAN 98], [SAN 98b]). In their paper on mobile cryptography, they deal with the protection of mobile agents from possibly malicious hosts. Hence, the focus in their work is on the use of cryptographic techniques for securing mobile code. The security concerns posed by the mobile agent protection problem are code privacy (can a mobile agent conceal the program it wants to have executed?), code and execution integrity (can a mobile agent protect itself against tampering by a malicious host?) and computing with secrets in public (can a mobile agent remotely sign a document without disclosing the user's private key?). To address some of these concerns, cryptographic secure computation techniques can be used. We discuss this in more detail in Section 2.3, which is part of our survey on secure distributed computing protocols.

The structure of this paper is as follows. In the next section, we start with a survey of existing cryptographic solutions to the secure computing problem. In Section 3, we introduce and compare three possible ways to implement secure distributed computing, making use of both cryptographic techniques, and trusted parties. The comparison is based on a simple model of the trust and communication requirements for each of the solutions. In Section 4, we focus on an example application of secure distributed computing. More precisely, we will show how multi-party secure computations can be used to perform second price auctions and we will assess the incurred communication overhead. Finally, in Section 5, we summarize the main outcomes of this contribution.

2. Survey of SDC protocols

Various kinds of solutions for the secure distributed computing problem have been proposed in the literature (often using different terminology than the one used in this paper).

2.1 Using probabilistic encryption

One class of techniques to compute with encrypted data is based on *homomorphic probabilistic encryption*. An encryption technique is *probabilistic* if the same cleartext can encrypt to many different ciphertexts. To work with encrypted bits, probabilistic encryption is essential, otherwise only two ciphertexts (the encryption of a zero and the encryption of a one) would be possible, and crypt-analysis would be fairly simple. An encryption technique is *homomorphic* if it satisfies equations of the form $E(x \text{ } \mathbf{op} \text{ } y) = E(x) \text{ } \mathbf{op'} \text{ } E(y)$ for some operations $\mathbf{op}$ and $\mathbf{op'}$. A homomorphic encryption scheme allows operations to be performed on encrypted data, and hence can be used for secure circuit evaluation.

Abadi and Feigenbaum present a protocol for two-player secure circuit evaluation using a homomorphic probabilistic encryption scheme based on the Quadratic Residuosity Assumption (QRA) in [ABA 90]. This protocol allows A who has a secret function f and B who has secret data x to calculate $f(x)$ without revealing their secrets.

Let k be the product of two primes p and q, each congruent to 3 mod 4. An integer $a \in Z_k^*[+1]$ – the integers relatively prime to k with Jacobi symbol 1 – is a quadratic residue mod k if there exists an $x \in Z_k^*[+1]$ such that $a = x^2 \bmod k$. The QRA states that determining if an integer a is a quadratic residue mod k is a difficult problem if the factorization of k is unknown but is easy to solve if p and q are given.

If we encrypt a zero by a quadratic residue and a one by a quadratic nonresidue mod k, we can define the encryption of a bit b as

$$E_k(b) = (-1)^b \cdot r^2 \bmod k$$

with $r \in_R Z_k^* [+1]$ chosen at random. This probabilistic encryption scheme has two homomorphic properties that will come in handy in the protocol:

$$E_k(\bar{b}) = (-1) \cdot E_k(b) \bmod k$$

$$E_k(b_1 \oplus b_2) = E_k(b_1) \cdot E_k(b_2) \bmod k$$

B starts the protocol by choosing p and q and multiplying them to produce k. B sends k and the encryption of his data bits $E_k(x_1)$, ..., $E_k(x_n)$ to A. B keeps the factorization of k secret. A then starts evaluating her secret circuit. If she has to evaluate a NOT gate with input $E_k(b)$, she simply calculates $- E_k(b) \bmod k$. An XOR with inputs $E_k(b_1)$ and $E_k(b_2)$ is also easy to evaluate: A just takes $E_k(b_1) \cdot E_k(b_2) \bmod k$ as the output of the gate. To evaluate the AND of inputs $E_k(b_1)$ and $E_k(b_2)$, she needs B's help. A chooses two bits c_1 and c_2 at random and sends $E_k(b_1 \oplus c_1)$ and $E_k(b_2 \oplus c_2)$ to B. B decrypts the bits A just sent him as d_1 and d_2 (he can do so because he knows p and q) and sends the tuple

$$< E_k(d_1 \wedge d_2), E_k(d_1 \wedge \bar{d}_2), E_k(\bar{d}_1 \wedge d_2), E_k(\bar{d}_1 \wedge \bar{d}_2) >$$

to A. A takes the first element of this tuple as the output of the AND gate if she chose $c_1 = c_2 = 0$, the second if she chose $c_1 = 0$ and $c_2 = 1$, the third if she chose $c_1 = 1$ and $c_2 = 0$ and the last one if she chose $c_1 = c_2 = 1$. Proceeding this way from gate to gate, A ends with the encrypted result $E_k(f(x))$ and sends it for decryption to B.

Note the large amount of communication overhead in the protocol: for each AND gate to be evaluated, a large amount of communication is necessary. Concrete estimates of the communication overhead in a realistic example can be found in [NEV 00].

2.2 Protocols based on oblivious transfer

In [GOL 87], Goldreich, Micali and Wigderson present a two-party protocol for the problem of *combined oblivious transfer* which is equivalent to the problem of secure circuit evaluation. The setting is slightly different than in the previous protocol. Here, two parties A and B want to evaluate a publicly known boolean circuit. This circuit takes input from both A and B, but each party wants to keep his part of the data private. In contrast, in the previous protocol, the circuit was private to A, and the data was private to B. Recall from the introduction that these two settings are essentially equivalent: by making the publicly known circuit an universal circuit, it is still possible to hide functions instead of data.

The basic idea of the protocol we are about to describe is the following: A will evaluate the circuit, not on the actual bits, but on encodings of those bits. The encoding of the bits is known only to B. So A evaluates the circuit, but cannot make sense of intermediary results because she doesn't know the encoding. B knows the encoding but never gets to see the intermediary results. When the final result is announced by A (in encoded form), B will announce a decoding for this final result.

We give a more detailed description of the protocol. B assigns two random bit strings r_i^0 and r_i^1 to every wire i in the circuit, which represent an encoded 0 and 1 on that wire. This defines a mapping $\phi_i : r_i^b \mapsto b$ for every wire i. B also chooses a random bit string R that will allow A to check if a decryption key is correct. The general idea of the protocol is that, if b is the bit on wire i in the evaluation of the circuit for A's and B's secret inputs, A will only find out about and r_i^b will never get any information about $\phi_i (r_i^b)$ or $r_i^{\bar b}$. In other words, A evaluates the circuit with encoded data.

We use the notation $E(M, r)$ for a symmetric encryption function of the message M with secret key r. To encrypt a NOT-gate with input wire l and output wire o, B constructs a random permutation of the tuple

$$< E(R \cdot r_o^1, r_i^0), E(R \cdot r_o^0, r_i^1) >$$

where $\cdot$ denotes the concatenation of bit strings. To encrypt an AND-gate with input wires l and r and output wire o, B constructs a random permutation of the tuple

$$< E(R \cdot r_o^0, r_l^0 \oplus r_r^0), E(R \cdot r_o^0, r_l^0 \oplus r_r^1),$$
$$E(R \cdot r_o^0, r_l^1 \oplus r_r^0), E(R \cdot r_o^1, r_l^1 \oplus r_r^1) >$$

with $\oplus$ the bit-wise XOR. Any other binary port can be encrypted in an analogous way.

B sends the encryption of every gate in the circuit together with R, the encoding of his own input bits and the mapping ϕ_m of the output wire m to A. To perform the evaluation of the circuit on encoded data, A first needs encodings of all the input bits. For B's input bits, the encoding was sent to her, but since B doesn't know A's inputs, B can't send an encoding of them. Note that B can't send the encoding of both a 1 and a 0 on A's input wires either, because that would allow A to find out more than just the result of the circuit. The technique that is used to get the encoding of A's input to A is called *one-out-of-two oblivious transfer* ([EVE 85]).

This is a protocol that allows A to retrieve one of two data items from B in such a way that (1) A gets exactly the one of two items she chose and (2) B doesn't know which item A has got.

Thus, A and B execute a one-out-of-two oblivious bit string transfer (often referred to as $\binom{2}{1} -OT^k$) for each of A's input bits. This guarantees that A only obtains the encoding of her own input bits without releasing any information about her bits to B. A evaluates each gate by trying to decrypt every element of the tuple using the encoding of the bit on the input wire (or the XOR of two input bit encodings) as a key; she will only decrypt one of the elements successfully, thereby obtaining the encoded bit on the output wire. Note that she can verify if a decryption was correct by comparing the first bits of the decrypted string with R. Proceeding this way through the entire circuit, A obtains the encoding of the final output and applies ϕ_m to reveal the plain output bit.

Another protocol for two-party secure computation based on oblivious transfer is presented in [GOL 87b]. The basic idea in this protocol is to have the participants compute the circuit on data that is shared by the two parties using a technique known as *secret sharing*.

2.3 Autonomous protocols

The protocols discussed in the two previous subsections require more communication rounds than strictly necessary. The probabilistic encryption based protocol requires one communication round per AND-gate in the circuit. The oblivious transfer based protocol requires one communication round for performing the oblivious transfer of the input, and another for sending the encrypted circuit.

For protecting mobile code privacy and integrity, non-interactive (or autonomous) protocols are necessary ([SAN 98b]). The idea here is to realize a system where a host can execute an encrypted function without having to decrypt it. This, functions would be encrypted such that the resulting transformation can be implemented as a mobile program that will be executed on a remote host. The executing computer will be able to execute the program's instructions but will not be able to understand the function that the program implements. Having function and execution privacy immediately yields execution integrity: an adversary can not modify a program in a goal-oriented way. Modifying single bits of the encrypted program would disturb its correct execution, but it is very hard to produce a desired outcome.

It turns out to be possible to construct such autonomous solutions where the client sends (in one message) an encrypted function f, and it receives from the server an encrypted result $f(x)$ in such a way that f remains private to the client and x remains private to the server.

Various autonomous protocols have been proposed in the literature. Sander and Tschudin ([SAN 98, SAN 98b]) introduce a technique that allows for a fairly efficient evaluation of polynomials in a ring of integers modulo n using a homomorphic encryption scheme. They also show how an autonomous protocol could be realized using compositions of rational functions.

Sander and Tschudin emphasize in their paper that securing single functions is not sufficient. Consider for example the problem of implementing a digital signing primitive for mobile agents. Even if the real signature routine can be kept secret, still the whole (encrypted but operational) routine might be abused to sign arbitrary documents. Thus, the second task is to guarantee that cryptographic primitives are unremovably attached to the data to which they are supposed to be applied (the linking problem). The general idea behind the solution here is to compose the signature generating function s with the function f of which the output is to be signed. Crucial for the security of this scheme is the difficulty of an adversary to decompose the final function into its elements s and f. An outline of how this could be implemented using rational functions is given in [SAN 98b].

Loureiro and Molva ([LOU 99]) use a public key encryption system based on Goppa codes. Their protocol allows for the evaluation of functions describable by a matrix multiplication. Loureiro and Molva also show how any boolean circuit evaluation can be done by a matrix multiplication. However, the representation of a boolean circuit requires a *huge* matrix (for a circuit with l inputs, one of the dimensions of the matrix is 2^l). It remains an open problem whether more efficient representations of boolean circuits as matrices can be achieved.

Finally, two very recent papers also focus on autonomous protocols: Sander, Young and Yung ([YUN 00]) propose an autonomous protocol based on a new homomorphic encryption scheme, and Cachin, Camenisch, Kilian and Müller ([CAC 00]) start from an OT-based SDC protocol as in Section 2.2, and succeed in merging the two phases of this protocol into one.

It is worth emphasizing that, even though autonomous protocols use the minimal number of messages, they do not solve the communication overhead problem: even though there are only two messages exchanged, these messages are extremely large.

2.4 Using group-oriented cryptography

All the previous protocols concentrate on the two-party case: only two parties are involved in the secure computation process. It is clear that the multi-party case is even more interesting from an application-oriented point of view. The multi-party case has also received considerable interest in the literature.

In [FRA 96], Franklin and Haber propose a protocol that is somewhat similar to the protocol by Abadi and Feigenbaum ([ABA 90]), in the sense that this protocol too evaluates a boolean circuit on data encrypted with a homomorphic probabilistic encryption scheme. The major difference between the two protocols, however, is that this protocol allows any number of parties to participate in the secure computation, while Abadi and Feigenbaum's protocol is restricted to two parties.

To extend the idea of [ABA 90] to the multi-party case, we need an encryption scheme that allows anyone to encrypt, but needs the cooperation of all participants to decrypt. In a joint encryption scheme, all participants know the public key K_{pub} while each participant $P_1, ..., P_n$ has his own private key $K_1, ..., K_n$. Using the public key, anyone can create an encryption $E_S(m)$ of some message m, where $S \subseteq \{P_1, ..., P_n\}$, such that the private key of each participant in S is needed to decrypt. More formally, if D_i denotes the decryption with P_i's private key, the relation between encryption and decryption is given by

$$D_i(E_S(m)) = E_{S \setminus \{P_i\}}(m)$$

The plaintext m should be easily recoverable from $E_\emptyset(m)$.

In the joint encryption scheme used by Franklin and Haber, a bit b is encrypted as

$$E_S(b) = \left[g^r \bmod N, (-1)^b \left(\prod_{j \in S} g^{K_j} \right)^r \bmod N \right]$$

where $N = pq$, p and q are two primes such that $p \equiv q \bmod 4$, and $r \in_R Z_N$. The public key is given by $[N, g, g^{K_1} \bmod N, ..., g^{K_n} \bmod N]$. This scheme has some additional properties that are used in the protocol:

– *XOR-Homomorphic.* Anyone can compute a joint encryption of the XOR of two jointly encrypted bits. Indeed, if $E_S(b) = [\alpha, \beta]$ and $E_S(b') = [\alpha', \beta']$, then $E_S(b \oplus b') = [\alpha\alpha' \bmod N, \beta\beta' \bmod N]$.

– *Blindable.* Given an encrypted bit, anyone can create a random ciphertext that decrypts to the same bit. Indeed, if $E_S(b) = [\alpha, \beta]$ and $r \in_R Z_N$, then

$$\left[\alpha g^r \bmod N, \beta \left(\prod_{j \in S} g^{K_j} \right)^r \bmod N \right]$$

is a joint encryption of the same bit.

– *Witnessable.* Any participant can withdraw from a joint encryption by providing the other participants with a single value. Indeed, if $E_S(b) = [\alpha, \beta]$, it is easy to compute $D_i(E_S(b))$ from

$$W_i([\alpha, \beta]) = \alpha^{-K_i} \bmod N$$

First of all, the participants must agree on a value for N and g, choose a secret key K_i and broadcast $g^{K_i} \bmod N$ to form the public key. To start the actual protocol, each participant broadcasts a joint encryption of his input bits. For an XOR-gate, everyone simply applies the XOR-homomorphism. The encrypted output of a NOT-gate can be found by applying the XOR-homomorphism with a default encryption of a one, e.g. $[1, -1]$.

Again, it is the AND-gate that causes some trouble. Suppose the encrypted input bits for the AND-gate are $\hat{u} = E(u)$ and $\hat{v} = E(v)$. To compute a joint encryption $\hat{w} = E(w) = E(u \wedge v)$, they proceed as follows:

1. Each participant P_i chooses random bits b_i and c_i and broadcasts $\hat{b}_i = E(b_i)$ and $\hat{c}_i = E(c_i)$.

2. Each participant repeatedly applies the XOR-homomorphism to calculate $\hat{u}' = E(u') = E(u \oplus b_1 \oplus \ldots \oplus b_n)$ and $\hat{v}'\ E(v') = E(v \oplus c_1 \oplus \ldots \oplus c_n)$. Each participant broadcasts decryption witnesses $W_i(\hat{u}')$ and $W_i(\hat{v}')$.

3. Everyone can now decrypt $\hat{u}'$ and $\hat{v}'$. We have the following relation between $w' = u' \wedge v'$ and $w = u \wedge v$:

$$
\begin{aligned}
w' \ &= u' \wedge v' \\
&= (u \oplus b_1 \oplus \ldots \oplus b_n) \wedge (v \oplus c_1 \oplus \ldots \oplus c_n) \\
&= (u \wedge v) \quad \oplus \quad (u \wedge c_1) \quad \oplus \ldots \oplus \quad (u \wedge c_n) \\
& \quad \oplus \quad (b_1 \wedge c_1) \quad \oplus \ldots \oplus \quad (b_n \wedge c_1) \\
& \quad \vdots \qquad\qquad \vdots \qquad\qquad\qquad \vdots \\
& \quad \oplus \quad (b_1 \wedge c_n) \quad \oplus \ldots \oplus \quad (b_n \wedge c_n) \\
& \quad \oplus \quad \underbrace{(b_1 \wedge v)}_{w_1} \quad \oplus \ldots \oplus \quad \underbrace{(b_n \wedge v)}_{w_n} \\
&= (u \wedge v) \oplus w_1 \oplus \ldots \oplus w_n
\end{aligned}
$$

Each participant is able to compute a joint encryption of w_i: he knows b_i and c_i (he chose them himself) and he received encryptions $\hat{c}_j$ from the other participants, so he can compute $E(b_i \wedge c_j)$ as follows:

– If $b_i = 0$, then $b_i \wedge c_j = 0$, so any default encryption for a zero will do, e.g. [1, 1].
– If $b_i = 1$, then $b_i \wedge c_j = c_j$, so $\hat{c}_j$ is a valid substitution for $E(b_i \wedge c_j)$.

$E(u \wedge c_i)$ and $E(v \wedge b_i)$ can be computed in an analogous way. He uses the XOR-homomorphism to combine all these terms, blinds the result and broadcasts this as $\hat{w}_i$.

4. Each participant combines $\hat{w}'$ and $\hat{w}_j$ ($j = 1 \ldots n$), again using the XOR-homomorphism, to form $\hat{w} = E(w)$.

When all gates in the circuit have been evaluated, every participant has a joint encryption of the output bits. Finally, all participants broadcast decryption witnesses to reveal the output.

2.5 Other multi-party protocols

Chaum, Damgård and van de Graaf present a multi-party protocol in [CHA 87] that starts with the truth tables of every gate in the circuit. Each player in turn receives a "scrambled" version of the truth tables from the previous player, transforms the truth tables by adding his own encryptions and permutations, commits to his encryptions and sends these transformed truth tables to the next player. When the last player finished his transformation, all players evaluate the scrambled circuit by selecting the appropriate row from the truth tables.

Even information-theoretically secure multi-party computation can be achieved (as opposed to only computationallly secure). A possible realisation is discussed in [CRA 99].

The communication overhead for multi-party protocols is even more serious than that for the two-party protocols.

3. Trust versus communication overhead

In this section, the different options for implementing secure distributed computation are discussed. It will be shown that there is a trade-off between trust and communication overhead in secure computations. If all participants are distrustful of each other, the secure computation can be performed using protocols surveyed in the previous section with a prohibitive huge amount of communication. However, if a TTP is involved, the communication overhead can be made minimal.

Recall from Section 1 that f is a publicly known function taking n inputs. Assume that there are n distrustful participants $p_1, \ldots, p_n$, each holding one private input x_i. The n participants want to compute the value of $f(x_1, \ldots, x_n)$ without leading information of their private inputs to the other participants.

To compare the trust requirements of the different approaches, we use the following simple trust model. We say a participant *trusts* an execution site if it believes that:

– the execution site will correctly execute any code sent to it by the participant;
– the execution site will correctly (i.e. as expected by the participant) handle any data sent to it by the participant.

It also implies that the execution site will maintain the privacy of the data or the code if this is expected by the participant. If p trusts E, we denote this as shown in Figure 1.

Figure 1. Notation for "p trusts E"

To compare bandwidth requirements (for communication overhead), we make the following simple distinction. *High* bandwidth is required to execute a SDC protocol. *Low* bandwidth suffices to transmit data or agent code. We assume low bandwidth communication is available between any two sites. If high bandwidth communication is possible between E_i and E_j, we denote this as shown in Figure 2.

Figure 2. *Notation for high bandwidth connection between E_i and E_j*

To see that this simple two-valued model of bandwidth requirements is sufficient for our case, we refer the reader to [NEV 00]. In that paper, a case study investigating the communication overhead for a so-called *Secret Query Database* is given. In this application, A has a query q and B owns a database with records x. The Secret Query Database allows them to cooperate in such a way that they can compute $q(x)$ while A preserves the secrecy of q and B preserves that of x. The communication overhead to solve this concrete case with SDC protocols is in the order of magnitude of 100 megabytes. On the other hand, sending just the query data, or sending an agent containing the query requires only a few kilobytes of communication. The large difference in amount of communication shows that our simplified model of high and low bandwidth requirements is realistic.

Based on these simple models of communication and trust, we compare the three options for implementing secure distributed computations.

3.1 A trusted third party

The first, perhaps most straightforward option, is to use a globally trusted third party. Every p_i sends its private input x_i to the TTP who will compute $f(x_1, \ldots, x_n)$ and disseminate the result to the participants p_i, $i = 1..n$.

Of course, before sending its private data to the TTP, every p_i must first authenticate the TTP, and then send x_i through a safe channel. This can be accomplished via conventional cryptographic techniques.

It is clear that this approach has a very low communication overhead: the data is only sent once to the TTP; later, every participant receives the result of the computation. However, every participant should unconditionally trust the TTP. For the case of 4 participants, the situation is as shown in Figure 3.

It is not clear whether n distrustful participants will easily agree on one single trustworthy execution site. This requirement of one single globally trusted execution site is the main disadvantage of this approach.

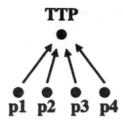

Figure 3. Situation with 4 participants and a TTP

3.2 Cryptographic secure distributed computing

The second option is the use of cryptographic techniques (as surveyed in Section 2) that make the use of a TTP superfluous.

The trust requirements are really minimal: every participant p_i trusts its own execution site E_i, and expects that the other participants provide correct values for their own inputs.

Although this option is very attractive, it should be clear from the previous sections and from [NEV 00] that the communication overhead is far too high to be practically useful in a general networked environment. High bandwidth is required between all of the participants. For the case of 4 participants, the situation can be summarized as shown in Figure 4.

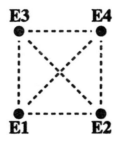

Figure 4. Situation with 4 participants without a TTP

3.3 A virtual trusted third party

Finally, our solution tries to combine the two previous options: the communication overhead of SDC-techniques are remedied by introducing semi-trusted execution sites and mobile agents.

In this approach, every participant p_i sends its representative, agent a_i, to a trusted execution site *Ej*. The agent contains a copy of the private data x_i and is capable of running a SDC-protocol.

It is allowed that different participants send their agents to different sites; the only restriction being that the sites should be located closely to each other, i.e. should have high bandwidth communication between them.

Of course, every execution site needs a mechanism to safely download an agent. However, that can be easily accomplished through conventional cryptographic techniques.

The amount of large distance communication is moderate: every participant sends its agent to a remote site, and receives the result from its agent. The agents use a SDC-protocol, which unfortunately involves a high communication overhead. However, since the agents are executing on sites that are near each other, the overhead of the SDC-protocol is acceptable. For a situation with 4 participants, we could have the situation as depicted in Figure 5.

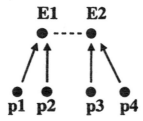

Figure 5. *Situation with 4 participants and a virtual trusted third party*

No high bandwidth communication between the participants is necessary, and there is no longer a need for one single trusted execution site. p_1, for example, does not need to trust site E_2. The agents that participate in the secure computation are protected against malicious behaviour of other (non-trusted) execution sites by the SDC-protocols. That is sufficient to make this approach work.

Moreover, in contrast with the approach where one uses an unconditionally trusted third party, the trusted sites are not involved directly. They simply offer a secure execution platform: the trusted hosts do *not* have to know the protocol used between the agents. In other words, the combination of mobile agent technology and secure distributed computing protocols makes it possible to use *generic* trusted third parties that, by offering a secure execution platform, can act as trusted third party for a wide variety of protocols in a uniform way.

Finally, the question remains whether it is realistic to assume that participants can find execution sites that are close enough to each other. Given the fact however that these execution sites can be *generic*, we believe that providing such execution sites could be a commercial occupation. Various deployment strategies are possible. Several service providers, each administering a set of geographically dispersed "secure hosts", can propose their subscribers an appropriate site for the secure computation. The site is chosen to be in the neighbourhood of a secure site of the other service providers involved. Another approach is to have execution parks, offering high bandwidth communication facilities, where companies can install their proprietary "secure site". The park itself could be managed by a commercial or government agency.

4. Case study: second price auctions

In this example we consider the case of a second price auction, where there is one item for sale and there are n bidders. The item will only be sold if the bid of one participant is strictly higher than the other bids. In all other cases there is no winner. The clearing price is the second highest bid. The requirements for this type of auction are the following:

- if there is no winner, do not reveal anything;
- if there is a winner:
  - reveal the identity of the highest bidder, but hide the highest bid;
  - reveal the 2nd highest bid, but hid the identity of the 2nd highest bidder;
  - do not reveal any other information.

Our goal is to estimate the communication overhead of an implementation of secure distributed second price auctions with the protocol proposed by Franklin and Haber (Section 2.4). The auction is designed as a boolean circuit and the communication overhead for secure circuit evaluation is estimated. The communication overhead is determined by the following steps in the protocol:

- broadcast of the encrypted input bit of each participant;
- evaluation of an AND gate:
  - broadcast of the encrypted bits $E(b_i)$, $E(c_i)$;
  - broadcast of the decryption witnesses $W_i(\hat{u}')$, $W_i(\hat{v}')$;
  - broadcast of the blinded $\hat{w}_i$;
- broadcast of the output decryption witnesses.

The associated communication overhead is:

- $2 \cdot |N| \cdot in_i \cdot n$ for the broadcast of the input bits;
- $8 \cdot |N| \cdot n$ for the evaluation of an AND gate;
- $|N| \cdot out \cdot n$ for the decryption broadcast.

where $|N|$ is the length of N in bits, which is the same as the number of bits needed to represent an element of Z_N^*, in_i is the number of input bits of participant i, n is the number of participants and out is the number of output bits of the circuit. In

Table 1. Network overhead of secure second price auctions

n	4	16	32
Overhead (MB)	15	1000	8000

order to estimate the communication overhead, we need to be able to determine the number of AND gates in the boolean circuit (remember that each OR gate can be implemented with AND and NOT gates).

For three participants X, Y and Z, the boolean circuit is shown in Figure 6. The inputs to the circuit are 32-bit bids. The output is the identity of the winner, represented by the bits $R1$ and $R0$ ($R1R0 = 00$ no winner, 01 winner is X, 10 winner is Y, 11 winner is Z), and the clearing price. If there is no winner, the clearing price is set to zero. To determine the winner, the circuit uses three comparators and a number of AND and OR gates. To determine the clearing price, four multiplexers are used. Consider the situation where X makes the highest bid. In this case $G1 \wedge G2 = 1$, $L1 \wedge G3 = 0$, $L2 \wedge L3 = 0$ and $R1R0 = 10$, so the second input to the final multiplexer will be chosen. The input on this line is determined by the bids made by Y and Z. If $Y > Z$ then $G3 = 1$ and Y will be selected as the clearing price. In the other cases ($Y < Z$ or $Y = Z$) Z will be the clearing price.

For $n > 3$ participants, the circuit changes as follows. The number of comparators needed is now $\binom{n}{2} = n \cdot (n-1)/2$ and each comparator has 434 AND gates. The final multiplexer will need to distinguish between $n + 1$ different cases, i.e. n possible winners or no winner at all. The other n multiplexers are there to select the clearing price out of $n - 1$ bids when there is a winner. The number of AND gates needed for each multiplexer as a function of the number of inputs m is shown in Figure 7. Besides the comparators and the multiplexers, some additional AND and OR gates are needed. However, the number of these gates is negligible compared to the number of gates needed for the comparators and multiplexers. In summary, the circuit has a total gate complexity of $O(n^2)$.

The results of estimating the communication overhead for this circuit as a function of the number of participants n are summarized in Table 1.[1] Franklin and Haber's protocol is linear in the number of broadcasts, so the total message complexity is $O(n^3)$. However, it must be noted that this only holds on a network with broadcast or multicast functionality, such that the communication overhead of sending a message to all participants is the same as that of sending a message to a single participant. In absence of such infrastructure, the total message complexity is $O(n^4)$.

The figures in Table 1 show that the mobile agent approach for implementing SDC protocols makes sense: the communication overhead is feasible on a LAN, but prohibitive over the Internet.

5. Conclusion

This paper shows how the use of semi-trusted hosts and mobile agents can provide for a trade-off between communication overhead and trust in secure distributed

---

1. We choose lM to be 1024 bits.

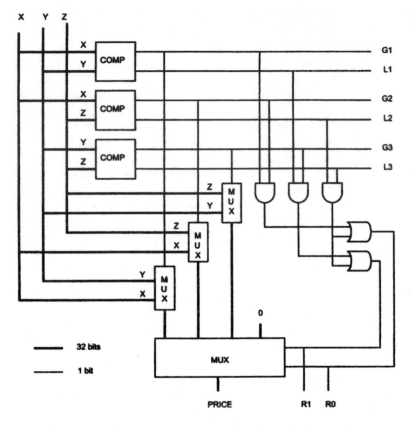

Figure 6. *Boolean circuit implementation of second price auctions*

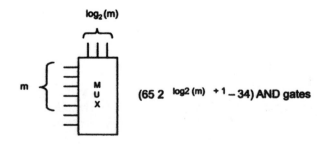

Figure 7. *Number of AND gates needed in a multiplexer*

computing. There is no need for one generally trusted site, nor does the program code have to be endorsed by all participants. The trusted execution sites are generic and can be used for a wide variety of applications. The communication overhead of secure distributed computing protocols is no longer prohibitive for their use since the execution sites are located closely to each other.

REFERENCES

[ABA 90] M. ABADI, J. FEIGENBAUM, "Secure Circuit Evaluation, a Protocol Based on Hiding Information from an Oracle", *Journal of Cryptology*, 2(1), p. 1–12, 1990.

[CAC 00] C. CACHIN, J. CAMENISCH, J. KILIAN, J. MÜLLER, "One Round Secure Computation and Secure Autonomous Mobile Agents", submitted to *ICALP 2000*.

[CHA 87] D. CHAUM, I. DAMGÅRD, J. VAN DE GRAAF, "Multiparty Computations Ensuring Privacy of Each Party's Input and Correctness of the Result", in *Advances in Cryptology – CRYPTO '87 Proceedings* (Lecture Notes in Computer Science, Vol. 293), ed. C. Pomerance, p. 87–119, Springer-Verlag, New York, 1988.

[CHO 95] B. CHOR, O. GOLDREICH, E. KUSHILEVITZ, M. SUDAN, "Private Information Retrieval", *Proceedings of 36th IEEE Conference on the Foundations of Computer Science (FOCS)*, p. 41–50, 1995.

[CRA 99] R. CRAMER, "An Introduction to Secure Computation", in *LNCS 1561*, p. 16–62, 1999.

[EVE 85] S. EVEN, O. GOLDREICH, A. LEMPEL, "A Randomized Protocol for Signing Contracts", *Communications of the ACM*, vol. 28, 1985, p. 637–647.

[FRA 93] M. FRANKLIN, "Complexity and Security of Distributed Protocols", Ph.D. thesis, Computer Science Department of Columbia University, New York, 1993.

[FRA 96] M. FRANKLIN AND S. HABER, "Joint Encryption and Message-efficient Secure Computation", *Journal of Cryptology*, 9(4), p. 217–232, Autumn 1996.

[GOL 87] O. GOLDREICH, S. MICALI, A. WIGDERSON, "How to Play any Mental Game", *Proceedings of 19th ACM Symposium on Theory of Computing (STOC)*, p. 218–229, 1987.

[GOL 87b] O. GOLDREICH, R. VAINISH, "How to Solve any Protocol Problem: an Efficiency Improvement", *Proceedings of Crypto '87, LNCS 293*, p. 73–86, Springer-Verlag, 1987.

[LOU 99] S. LOUREIRO, R. MOLVA, "Privacy for Mobile Code", *Proceedings of the Workshop on Distributed Object Security, OOPSLA '99*, p. 37–42.

[NEV 00] G. NEVEN, F. PIESSENS, B. DE DECKER, "On the Practical Feasibility of Secure Distributed Computing: a Case Study", *Information Security for Global Information Infrastructures* (S. Qing and J. Eloff, eds.), Kluwer Academic Publishers, 2000, p. 361–370.

[NIS 99] N. NISAN, "Algorithms for Selfish Agents", *Proceedings of the 16th Annual Symposium on Theoretical Aspects of Computer Science*, Trier, Germany, March 1999, p. 1–15.

[SAN 98] T. SANDER, C. TSCHUDIN, "On Software Protection via Function Hiding", *Proceedings of the Second Workshop on Information Hiding,* Portland, Oregon, USA, April 1998.

[SAN 98b] T. SANDER, C. TSCHUDIN, "Towards Mobile Cryptography", *Proceedings of the 1998 IEEE Symposium on Security and Privacy,* Oakland, California, May 1998.

[YUN 00] T. SANDER, A. YOUNG, M. YUNG, "Non-Interactive CryptoComputing for NC^1", preprint.

Chapter 2

Network domain agency for QoS management in OSPF configured networks

Farag Sallabi and Ahmed Karmouch

Multimedia Information and Mobile Agents Research Laboratory, School of Information Technology and Engineering, University of Ottawa, Canada

1. Introduction

Recent multimedia applications require stringent values of QoS. Introducing high quality video and audio signals has made it difficult for the existing infrastructure of the Internet to cope with those multimedia applications. Furthermore, recent applications are associated with user interactions, and the ability to browse different scenarios at the same time. In fact, these services made the researchers look for other solutions and even make changes to the existing Internet infrastructure. The problem of the QoS in the Internet requires a close cooperation between different protocols and components. For example, in the network there should be cooperation between the session setup protocols and the routing protocols to establish a path that can handle the QoS request. This also requires the cooperation of admission control, packet classifiers and packet schedulers.

The literature on QoS contains many proposals for fulfilling the requirements of real-time applications. All proposals agree that the best way to guarantee QoS is to provide some sort of resource reservations in the network elements. Resource reservations could be done immediately or in advance, to reflect human life in arranging and monitoring their activity schedules. To increase the admission rate of real-time applications several proposals for improving the routing protocols, to be QoS sensitive, have been suggested.

In previous work [SAL 99b] we have developed end-to-end resource reservations architecture. The architecture treats the Internet as a collection of autonomous systems that are configured and operated by the OSPF protocol. In OSPF, each autonomous system can be divided into areas interconnected by a backbone area. Therefore, we have adopted this configuration scheme and introduced a domain agency in each area (domain). In fact, if these areas are selected properly we can manage their network resources effectively, and obtain a better scalability for admitting immediate and advance resource reservations. We state in this paper a framework, which outlines the operation of the domain agencies in the OSPF configured networks.

In this paper, we present architecture for network domain agency, which we have designed to handle the resource reservations and QoS routing. We have used a different approach for QoS routing, where the underlying routing protocol in the network works as the background protocol for routing background packets. The QoS routing is invoked only when there is need for resource reservations. Resources in the domain are partitioned between best-effort service and QoS service.

The rest of this paper is organized as follows. In Section 2, we highlight the benefits of using agent technology in QoS managements. In Section 3, we state the mechanism of the end-to-end resource reservations and resource managements. The domain agency architecture is presented in Section 4. Section 5 explains the operation of the domain agencies in response to AS changes. In Section 6, we discuss the implementation directions. Finally, we conclude the paper in Section 7.

2. Supporting QoS protocols with software agents

Quality of service provisioning in computer networks requires at least two processes, QoS negotiation process and resources setup process. The two processes require the exchange of messages between end systems (which may require the presence of the user), and between end systems and the network elements. The negotiation process continues in exchanging messages until a final decision is made. In fact, this process may take a long time to finish, depending on the resources' availability. If we account for the fast expansion of the Internet, the new sophisticated multimedia applications and the fast growing number of Internet users, we realize that there is a non-negligible amount of traffic, due to QoS negotiation and resources setup, which puts extra load on the network resources. The alternative way of using classical QoS negotiation and resource setup is to use Intelligent Agents (IA), which work on behalf of the user or another entity to complete a certain task. In the negotiation process agents are provided with the necessary information that enable them to act intelligently and accomplish their job without the intervention of users.

There are two types of agents; static agents and mobile agents. Static agents accomplish their task, in the host node, without the need to move to another node, while mobile agents have the ability to move from node to another node in order to cooperate with other agents and/or software to complete a certain job. In fact, if those types of agents are used effectively in the network domains, the QoS negotiation and resources setup processes could be made much easier, scalable, flexible and fast. Following is a short description of the agent technology.

Pattie Maes, one of the pioneers of agent research, defined agents as "Autonomous agents are computational systems that inhabit some complex dynamic environment, sense and act autonomously in this environment, and by doing so realize a set of goals or tasks for which they are designed" [MAE 96]. The distributed nature of the Internet has contributed in the addition of mobility to be one of the agent behaviours. In fact, agents should be able to communicate and cooperate with local and remote agents and should be able to migrate to remote hosts to operate closer to physical data stores [FAL 98].

Nwana defined mobile agents as "Computational software processes capable of roaming wide area networks (WANs) such as the www, interacting with foreign hosts, gathering information on behalf of its owners and coming 'back home' having performed the duties set by its user" [NWA 96]. These duties may range from a flight reservation to managing a telecommunication network. The idea of a self controlled program execution near the data source has been proposed as the next wave to replace the client-server paradigm as a better, more efficient and flexible mode of communication [PHA 98]. Agents exhibit some properties that make them a best candidate for being used in the QoS management in computer networks. Those properties have been defined by Nwana [NWA 96] that include but are not limited to the following:

– Autonomy: The ability to operate without human intervention;
– Cooperation: The ability to interact with other agents and or human via some communication language;
– Learning: The ability to learn as they interact with external environment.

3. Resource allocation in network domains

Figure 1 exhibits a sample network with network domains. The resource management mechanism relies on finding the best path from the client to the server, then allocating the necessary resources. The network domains are the same as OSPF [MOY 98] domains/areas. In the OSPF, each router in a certain domain maintains an identical database describing the domain's topology. Therefore, any router in the domain can calculate the routing table of any other router. This property has been taken into account in designing the QoS routing agent (described later). The following sections describe the resource reservations process from the receiver to the sender.

3.1 Negotiation phase

The negotiation process is further divided into two stages. The first stage is between the receiver and the sender, and the second stage is between the end systems (receiver and sender) and the domain agencies in the network domains. This process includes the following types of messages:

– *Request Message:* This message is carried by the presentation agent, which takes it to the server for negotiation. It contains the user requirements.

– *Path Message:* This message is carried by the presentation agent and is sent from the sender to the domain agency in the sender domain. The message carries the traffic and time specifications, in addition to other information. After the agency finishes its job, it forwards the message to the next domain agency in the way to the receiver until it gets to the receiver.

– *Reserve Message:* The receiver sends this message just after it receives the *Path Message.* The message follows the reverse path of the *Path Message and* carries the same information. The *Reserve Message* is treated as a reserve confirmation message.

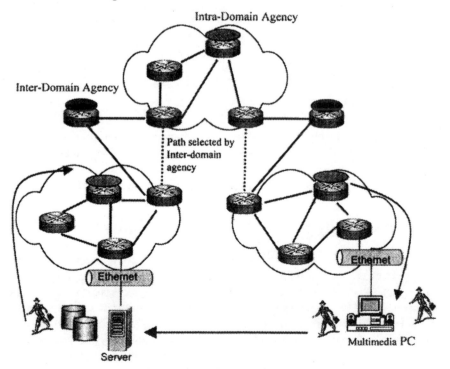

Figure 1. Sample network domains

3.1.1 Receiver-sender negotiation

The negotiation process starts as soon as the user involves his/her request. The *User Request* launches a Presentation Agent (PA), which receives the user information and works on behalf of the user to finish the negotiation process. Before the PA goes to the sender(s) site, it consults the admission control unit in the receiver side to make sure that the system can handle the user requirements. After accomplishing this step the PA travels to the sender(s) carrying the information that is common between the user requirements and the available system resources. When the PA arrives at the sender side, it negotiates with the sender the requested scenario and the QoS parameters. The result of the negotiation will be as follows:

$$PA_{load} = User\ Requirements \cap Available\ System\ Resources \qquad [1]$$

$$Scenario\ (Tspec) = [Scenario_i\ (Tspec) \mid Scenario_i\ (Tspec) \leq PA_{load}] \qquad [2]$$
$$i = 1, ..., n.$$

When $Scenario_1$ *(Tspec)* has the highest QoS parameters.

$$Scenario\ (Tspec,\ Timespec) = [(Scenario\ (Tspec) + User\ start\ time) \mid$$
$$(Scenario\ (Tspec) + User\ start\ time) \leq Available\ Server\ Resources\ at$$
$$specified\ time\ duration] \qquad [3]$$

Equation [1] is a result of applying admission control at the receiver side; the result should satisfy the user requirements in terms of QoS and presentation time, otherwise the user is instructed to select another parameter. Then in equation [2] the sender retrieves, from the database, the scenario that has the best traffic specifications (Tspec) and suits the PA's requirements. Then in equation [3] the sender submits the selected scenario together with the start time (supplied by the PA) to the admission control in the sender to admit the new scenario. If the scenario is accepted, the sender starts negotiating with the network. If the scenario is not accepted, the sender re-negotiates with the PA alternative QoS parameters and starting time. If all alternatives are not accepted the PA returns carrying a reject message.

3.1.2 End system-network negotiation

After finishing the receiver-sender negotiation, the PA carries the *Path Message* to the domain agency in the sender domain. The *Path Message* carries the session's Tspec, time specification (Timespec) and session addresses. The Tspec is used by the routing agent and admission control in the Domain Agency (DA) to prevent over-reservation, and admission failure. The DA establishes a QoS route (according to the Tspec and Timespec) in the domain towards the receiver(s), and reserves resources temporarily. Then it forwards the *PA* to the next DA in the way to the receiver. This process continues until the *PA* reaches the receiver. At this time, the receiver sends a *Reserve Message* to the DA in its area. The DA handles the *Reserve Message* as a reserve confirmation, and makes the necessary updates in the reservation state. Then the *Reserve Message* is forwarded to the next DA towards the sender and sends confirmation back to the previous DA or receiver. The process is repeated until the *Reserve Message* reaches the sender.

3.2 Resource reservation phase

This phase is concerned with making the actual resource reservations in routers and end systems, and maintaining them. Each domain agency and end system checks its reservation state for due sessions. If there are sessions ready to start, the DA sends messages to the RSVP on routers in its domain that are involved in the session. The messages set some parameters in packet classifiers and packet schedulers to obtain the desired QoS. The domain DA is also responsible for maintaining the reserved resources by sending refresh messages at specified time intervals. In the end systems, a Resource Reservation Agent (RRA) sets parameters in its node to provide the due sessions with its requested QoS. It also sends a refresh message to the DA. This phase includes the following messages:

   – *Start Message:* Just before the playback of every session, the domain agency sends a start message to the routers involved in this session, to set some parameters in packet classifiers and packet schedulers.

   – *Refresh Message:* This message is sent from the receiver to maintain the reservations. The message is sent at certain time intervals to the domain agencies, and the agencies are responsible for sending refresh messages to the RSVP in the routers. Time-out of refresh messages will cause the domain agency to teardown the session.

   – *Session Teardown Message:* This message can be initiated by the sender or by the receiver. It ends a specific session and releases the resources used by this session.

3.3 Adaptation phase

Adaptation is concerned with adjusting reserved resources according to user interactions. The user can interactively participate in the presentation of the multimedia application, where he/she can use VCR like control buttons, or link to other scenarios from the current scenario; the user may also make changes to reserved resources not yet started. These interactions will certainly introduce a change in the reserved resources, either in QoS parameters or in time duration. Any user interaction is received by the RRA that makes the necessary calculations to find the change. Then the request is forwarded to the admission control in the receiver.

If the change is accepted, the receiver RRA sends the user interaction request to the sender RRA. This step is necessary to check first if the sender can handle the user interaction request. The sender RRA responds to the receiver RRA by either accept or reject. If accepted the receiver RRA informs the domain agencies with the change for adapting the reserved resources. If one or more network domains reject the interaction request, a best effort service could be used. This phase includes the following message:

– *User Interaction Message:* This message is sent during the active or the passive state of the session. If there is any user interaction, the action is first negotiated between the receiver and the sender. If it is accepted, it is forwarded by the receiver to the domain agencies to update the reserved resources. The interaction request may come at a time where all resources are exploited; at this time the best-effort service can be used.

4. Domain agency architecture

The resource reservations scheme described in the previous section relies on domain agencies, in network domains, to manage domain resources. The domain agency receives resource reservations requests from end systems and agencies in other domains. This architecture has been adopted for several reasons. The first reason is to provide better advance and immediate resource reservation scalability. In previous resource reservation architectures [BRA 97], every network element is responsible for reserving resources and maintaining them. Due to the high volume of resource reservations, this would overload the routers and affect their performance. The problem is getting even worse when advance reservations have been permitted for users, which encourage many users to reserve resources in advance. Therefore, by this architecture the agency will take care of resource reservations and maintain their states. The other reason for using this architecture is the QoS route calculations. Sessions that need specific QoS should send their requests to the domain agency, which calculates the best-path in the domain and reserves the necessary resources in this path. The traditional routing protocol works in its normal way for forwarding background traffic. This will relieve the QoS routing protocol from finding the QoS routes on demand, as other proposals suggest [APO 99b, CRA 98, SHA 98, WHA 96, ZHA 97]. The other reason for using domain agencies is that the domain agency takes care of any problems in the domain and hides them from the end systems. In the example of route failure, the domain agency handles it without informing the end systems. The agency takes care of refresh messages as well, which allow us to control the flow of the refresh messages from end systems to the domain agencies and from the domain agencies to the routers within the domain. Figure 2 shows the components of the domain agency. The agency should reside in any router except the area boarder routers and there should be a backup domain agency.

4.1 Domain Resource Reservation Agent

The *Domain Resource Reservation Agent (DRRA)* works as a coordinator in the domain, where it receives the *Path Message* from the PA, and then it instructs the DACA to provide the available resources. Then it submits the traffic and time specifications together with the available resources to the QoSRA. The DRRA has also interface to the RSVP, through which it receives and sends the messages to RSVP in routers to setup QoS parameters and maintain reservations. The DRRA is also responsible for updating the reservations state agent. This update can be either adding accepted sessions or removing finished sessions. If the DRRA receives any *User Interaction Message* it forwards it to the DACA.

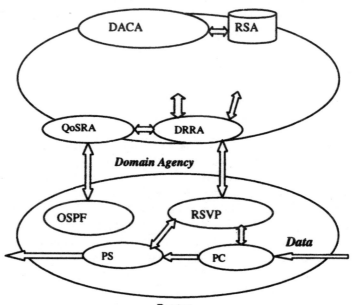

DACA: Domain admission control agent
DRRA: Open shortest path first
OSPF: Open shortest path first
PC: Packet classifier
PS: Packet scheduler
QoSRA: Quality of service routing agent
RSA: Reservations state agent
RSVP: Resource reservations protocol

Figure 2. *Domain agency architecture*

4.2 Quality of Service Routing Agent

The *Quality of Service Routing Agent (QoSRA)* has been introduced to find the best-path that satisfies the QoS needs of the multimedia applications [SAL 99a]. The QoSRA runs a modified dijkstra algorithm based on domain topology obtained from the OSPF, and the available link resources obtained from the domain admission control agent. The QoSRA is based on the OSPF property, which enables any router to calculate the routing table of any other router. Therefore, all resource reservations requests are sent to the domain agency that instructs the QoSRA to construct the QoS path. In fact, this is a new departure from other proposals [APO 99a, CRA 98, SHA 98, WHA 96, ZHA 97] that suggest enhancing the existing routing protocols to support QoS. In these proposals the calculation is done by every router, which overloads the routers and makes the advance reservations almost impossible, because of the scalability problem, and overhead result from calculating advance routes.

Two QoS routing calculation approaches are considered in the literature so far [APO 99b]. The first one is the on-demand calculation. In this approach the QoS path is calculated at the arrival of every request (requests that demand constraint QoS), this calculation is done by every router receiving the request. This approach is not scalable in the fast growing Internet. This situation is even worse if the QoS requests include advance reservations. The second approach is to pre-compute paths to all destinations for each node. Then at the arrival of every request, the path that satisfies the requested QoS is selected to forward the session's packets. This approach has many disadvantages. The QoS paths should be frequently pre-computed to reflect the current available resources. For every arrived request, the QoS paths need to be searched for the suitable path, which adds extra processing. Due to the high demand on QoS, the pre-computed paths are outdated and do not reflect the currently existing resources. Furthermore, the pre-computed paths are useless for advance reservations.

4.3 Domain Admission Control Agent

The *Domain Admission Control Agent (DACA)* has two functions [SAL 99a, SAL 99c]. The first one is to find the available link resources of all links in the domain at specified time intervals and submit this information to the DRRA (to be used by the QoSRA). The second function is to receive user interaction requests from the DRRA and apply admission control algorithm.

The idea of monitoring the available link capacities is as follows. At the beginning the QoS routing agent receives the domain topology from the OSPF and submits it to the DRRA. The DRRA inquires about each link capacity and submits the complete information to the reservations state agent. At each resource reservation request, the admission control finds the available link resources, for the specified time intervals, and then gives this information to the DRRA. The QoS routing agent constructs the best-path for this request and submits the resulting path to the DRRA. The DRRA reserves the resources temporarily until it receives a confirmation message from the receiver.

4.4 Reservations State Agent

The *Reservations State Agent (RSA)* stores all accepted reservation states. For better performance, the RSA is divided into two parts. The first part keeps track of link reservation states, while the second part keeps track of aggregated link reservation states. Figure 3 shows a snapshot of resource reservations made in a single link. The numbers indicate session's id, while α indicates a new session asking for admission. As we can see in the figure the duration as well as the bandwidth of each session is clearly defined. Therefore, if we aggregate the reservation state of every link, we get a summary database as described in Figure 4. This summary database enables the domain admission control agent to get quick values of the available link resources.

The RSA periodically checks the reservation database for due sessions; if there are any sessions ready to start, it sends the session's traffic specification to the DRRA that forwards them to the RSVP in the routers. The RSVP then set some parameters in the packet classifier and packet scheduler to achieve the requested QoS. The time of the playback of the scenario is controlled by the RSA, where it sends a start and finish message to the DRRA.

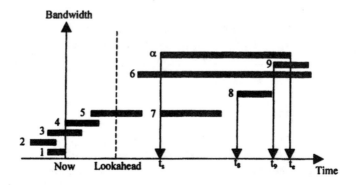

Figure 3. Snapshot of reservation state of a single link

5. Domain agency communications protocol

In previous sections, we saw how domain agencies receive and send QoS management messages. In this section we state a communications protocol that enables domain agencies to exchange messages about domain changes and

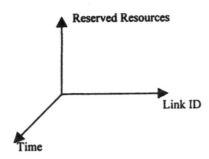

Figure 4. *Dimensions of the reservation database*

policies. In fact, as the OSPF domains change, by the network administrators, the domain agencies should react quickly to these changes in order not to lose any information and to adapt to the new configuration. The following section describes the communications protocol.

5.1 Interactions between domain agencies

Assume that there is one AS and one DA that manages the whole AS resources. As soon as the AS is split into areas, the QoSRA will be aware of this separation from the OSPF. The QoSRA informs the DRRA with the new AS configuration. From this configuration, the DRRA knows the addresses and number of the new areas. Then it duplicates and sends one DA to each new area including the backbone (the backbone is an area with address 0.0.0.0). Each DA resides in a router other than the area border router.

After each DA settles in its domain, it needs to communicate back with the parent DA to get its information. The DRRA in the child DA gets the topology data from the QoSRA, and sends this information to the DRRA in the parent DA. The DRRA in the parent DA gets the information related to each child DA from the RSA, according to topology database of each child DA, and then sends it back to its child DA. At the same time the parent DA constructs addresses table for the new DAs, and sends this table to every DA. This step is important to enable the DAs to communicate with each other. The parent DA should fall into one of the new areas including the backbone. Figure 5 shows the parent DA in the backbone and other DAs communicate with it to fetch their information.

There are five packet types that are used to convey the information between domain agencies.

– *Database description packet:* This packet is sent by the child DA to the parent DA. It carries the topology database of the child's domain.

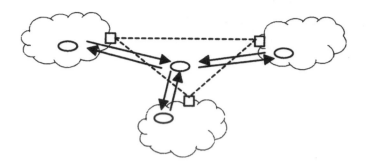

Figure 5. Interactions between domain agencies and the parent DA

– *Reservations State:* For each child DA, the parent DA collects the reservations state from the RSA and sends them back to the child DA. The message contains also the addresses table of the other DAs.

– *Policy and Service Agreement:* Domain agencies need to cooperate with each other to set up policies that each DA should follow. DA could also request another DA to reserve a certain path with certain bandwidth.

– *Hello Packet:* Every DA should send Hello packets to its neighboring DAs, to maintain relationships and check reachability. The DAs should also send Hello packets to end systems, to identify themselves and maintain reachability.

– *Acknowledgment Packet:* This packet is sent between the parent DA and the child DA. It has different interpretation as shown in Figure 6.

5.2 Status of current (immediate and advance) reservations

Because of the currently running sessions and the current reservations, the domain partitioning and the exchange of configuration messages should not create any disruptions for the currently running sessions, and the current reservations should be valid for the new configuration as well, otherwise the architecture would be useless. Figure 7 [MOY 98] presents an AS before and after partition. In Figure 7a all network elements belong to the same area (AS). In this configuration, only one DA manages the AS resources. If this AS is split into several areas as shown in Figure 7b, the DA will react quickly and dispatch DA for each new area. All borders that have been introduced to create areas are virtual, the location and interfaces of the network elements have not been changed. Therefore, the current resources reservations will also be applicable for the new configuration, and the active sessions will not be affected.

6. Implementation directions and experience

As we have stated previously, the end-to-end resource reservations and QoS management relies largely on the performance of the domain agency. In the end systems, we can easily control the scalability issue and the performance of the systems, especially with the appearance of the new machines with powerful speeds and storage. So the main problem is still concentrated in the network resources, with the increasing number of users and the emergence of sophisticated multimedia applications. Therefore, in our domain agency architecture we focus on the performance of the individual agents and the scalability of the domain agency. For these reasons, before we start implementing the DA, we have tested the QoSRA and the DACA using simulations to verify their functionality. In the following sections, we comment on the simulations and the next section presents the implementation directions.

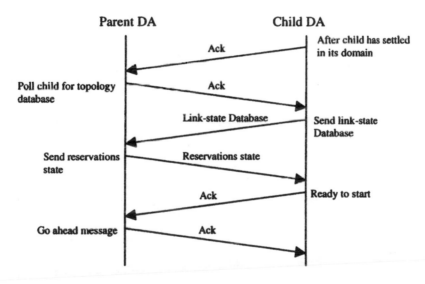

Figure 6. Parent DA and child DA setup

6.1 Simulation experience

We have performed simulations for single path admission control with and without user interactions [SAL 99c] and then applied the QoS routing algorithm on a certain network topology to get multiple paths (QoS paths) [SAL 99a], and then we compare the results of the simulations. The performance measures for the simulation was to test the advance and immediate rejection probabilities, the preemption probability and to compare the duration of the admitted advance and immediate flows. Table 1 shows the results of the single path simulation and Table 2 shows the results of the multiple paths simulation.

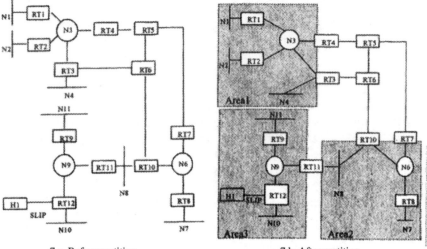

7.a. Before partition   7.b. After partition

Figure 7. OSPF AS domain

Table 1. Simulation results for single path

	Look-ahead Time			
	0	250	500	1000
Advance Rej. Prob.	.07	.07	.07	.07
Immediate Rej. Prob.	.42	.75	.78	.83
Preemption Prob.	.54	.13	.06	.02

Table 2. Simulation results for multiple paths

	Look-ahead Time			
	0	250	500	1000
Advance Rej. Prob.	.0	.0	.0	.0
Immediate Rej. Prob.	.0	.07	.1	. 125
Preemption Prob.	.56	.08	.04	.02

The offering load for single path admission control is 170%, while for multiple paths it is 200%. As we can see from the tables, there is improvement in the immediate and advance rejection probabilities in the multiple paths than in the single path. The preemption probability is also slightly improved and depends on the look-ahead time (the time that we account in admitting immediate flows). We noticed also that the duration of the admitted advance flows is significantly improved, i.e. the flows with long time duration have big chance to be admitted in the multiple paths than in the single path, see Figure 8 and Figure 9.

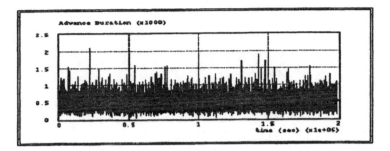

Figure 8. Admitted advance duration for single link

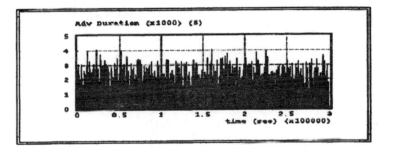

Figure 9. Admitted advance duration using QoS routing agent

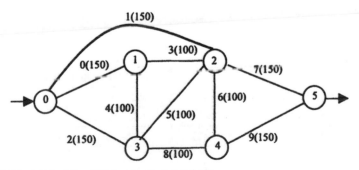

Where: X (Y) are linked Id and Link capacity, respectively

Figure 10. Simulated network topology

In addition, the average link utilization in multiple paths is high according to the QoS routing algorithm that we have used in calculating the best path [SAL 99a]. Figure 10 shows the network topology used in the simulation and Table 3 depicts the average utilization of each link.

6.2 Implementation directions

From the promising results obtained for the simulations, we moved to the implementation stage. As we have stated in previous sections, domain agencies need to communicate with each other, and agents in the same domain agency exchange messages with each other. Therefore, we have adopted the FIPA-OS (Foundation for Intelligent Physical Agents-Open Source) as an agent platform. FIPA-OS is a product from Nortel [NOR 00], which implements the FIPA specifications [FIP 96]. Figure 11 shows how the domain agency is mounted on top of the FIPA-OS. According to the software agent definition and properties, we have configured the domain agency with Agents that provide high performance, scalability and flexibility. Agents in the domain agency communicate with the software (OSPF and RSVP) using agent wrappers. The OSPF agent wrapper gets information from the OSPF protocol and submits them to the QoSRA. The RSVP agent wrapper relays commands from the DRRA to the RSVP protocol, and forwards messages from RSVP to DRRA.

Table 3. *Average link utilization*

Link Id	Look-ahead Time			
	0	250	500	1000
Link 0 Ave. Util.	.959	.984	.988	.990
Link 1 Ave. Util.	.369	.767	.782	.787
Link 2 Ave. Util.	.0	.0098	.015	.0279
Link 3 Ave. Util.	.968	.993	.995	.996
Link 4 Ave. Util.	.495	.542	.543	.543
Link 5 Ave. Util.	.830	.934	.940	.947
Link 6 Ave. Util.	.129	.646	.683	.705
Link 7 Ave. Util.	.956	.982	.987	.989
Link 8 Ave. Util.	.424	.540	.538	.545
Link 9 Ave. Util.	.367	.778	.8	.817

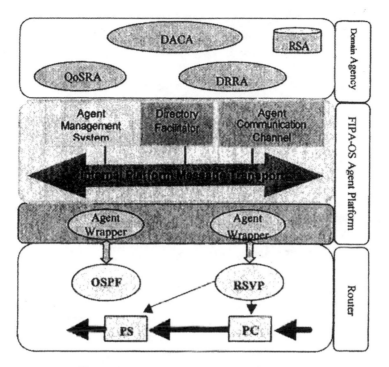

Figure 11. Using FIPA-OS as an agent platform

7. Conclusion

Network domains provide a better chance for network protocol developers to scale their protocols and get control of the resources in the domain. The OSPF routing protocol has the advantage of dividing an autonomous system into manageable areas, where each node in the domain knows only the nodes in its domain. We have considered this advantage in designing our resource reservation architecture and assigned a domain agency in every domain. In this paper, we have provided a framework, which enable domain agencies to react to area splitting and communicate with each other and with the end systems.

REFERENCES

[APO 99a] APOSTOLOPOULOS G., GUERIN R., KAMAT S., ORDA A., TRIPATHI S., "Intra-Domain QoS Routing in IP Networks: A Feasibility and Cost/Benefit Analysis", *Special Issue of IEEE Networks on Integrated and Differentiated Services for the Internet*, September 1999.

[APO 99b] APOSTOLOPOULOS G., GUERIN R., KAMAT S., ORDA A., PRZYGIENDA T., WILLIAMS D., "QoS Routing Mechanisms and OSPF Extensions", *RFC 2676, Experimental RFC, Internet Engineering Task Force*, August 1999.

[AUR 98] AURRECOECHEA C., CAMPBELL A., HAUW L., "A Survey of QoS Architectures", *ACM/Springer Verlag Multimedia Systems Journal, Special Issue on QoS Architecture*, Vol. 6, No. 3, p. 138–151, May 1998.

[BER 98] BERSON S., LINDELL R., BRADEN R., "An Architecture for Advance Reservations in the Internet", *Technical Report*, USC Information Sciences Institute, July 16, 1998.

[BRA 97] BRADEN R., ZHANG L., BERSON S., HERZOG S., JAMIN S., "Resource Reservation Protocol (RSVP) – Version 1 Functional Specification", *RFC 2205*, September 1997. ftp://ftp.ietf.org/internet-drafts-ietf-rvsp-spec-16.ps

[CLA 92] CLARK D., SHENKER S., ZHANG L., "Support Real-Time Applications in an Integrated Services Packet Network: Architecture and Mechanism", *Proceedings Of ACM SIGCOMM'92*, August 1992, p. 14–26.

[CRA 98] CRAWLEY E., NAIR R., RAJAGOPOLAN B., SANDICK H., "A Framework for QoS-Based Routing in the Internet", *RFC 2386*, August 1998.

[FAL 98] FALCHUK B., "A Platform for Mobile Agent-Based Data, Access, Retrieval, and Integration", *Ph.D. Thesis*, University of Ottawa, December 1998.

[FER 95] FERRARI D., GUPTA A., VERTRE G., "Distributed Advance Reservation of Real-Time Connections", *Fifth International Workshop on Network and Operating System Support for Digital Audio and Video*, Durham, NH, USA, April 19–21, 1995.

[FIP 96] http://www.fipa.org/, Established 1996.

[HAF 96] HAFID A., BOCKMANN G., DSSOULI R., "A Quality of Service Negotiation Approach with Future Reservation (NAFUR): A Detailed Study", *Technical Report*, University of Montreal, Montreal, 1996.

[MAE 95] MAES P., "Artificial Life meets Entertainment: Interacting with Lifelike Autonomous Agents", *Special Issue on New Horizons of Commercial and Industrial AI*, Vol. 38, No. 11, p. 108–114, *Communications of the ACM*, ACM Press, November 1995.

[MOY 98] MOY J., "OSFP Version 2", *RFC 2328*, April 1998.

[NOR 00] http://www.nortelnetworks.com/products/announcements/fipa/, 2000.

[NWA 96] NWANA H., "Software Agents: An Overview", *Knowledge Engineering Review*, Vol. 11, No. 3, p. 205–244, October/November 1996.

[PHA 98] PHAM V., KARMOUCH A., "Mobile Software Agents: An Overview", *IEEE Communications Magazine*, July 1998, p. 26–37.

[REI 94] REINHARDT W., "Advance Reservation of Network Resources for Multimedia Applications", *Proceedings of the Second International Workshop on Advanced Teleservices and High Speed Communication Architectures*, 26–28 September 1994, Heidelberg, Germany.

[SAL 99a] Sallabi F., Karmouch A., "Immediate and Advance Resource Reservations Architecture with Quality of Service Routing Agents", *Proceedings of Multimedia Modeling (MMM'99)*, October 4–6, 1999, Ottawa, Canada.

[SAL 99b] Sallabi F., Karmouch A., "New Resource Reservation Architecture with User Interactions", *Proceedings of IEEE Pacific Rim Conference on Communications, Computer and Signal Processing*, August 22–24, 1999, Victoria, BC, Canada.

[SAL 99c] Sallabi F., Karmouch A., "Resource Reservation Admission Control Algorithm with User Interactions", *Proceedings of the Globecom '99*, December 5–9, 1999, Rio de Janeiro, Brazil.

[SCH 97] Schelen O., Pink S., "Sharing Resources Through Advance Reservation Agents", *Proceedings of IFIP Fifth International Workshop on Quality of Service (IWQoS'97)*, New York, May 1997, p. 265–276.

[SHA 98] Shaikh A., Rexford J., Shin K., "Efficient Precomputation of Quality-of-Service Routes", *Proceedings of Workshop on Network and Operating Systems Support for Digital Audio and Video (NOSSDAV'98)*, July 1998, Cambridge, England, p. 15–27.

[WHA 96] Whang Z., Crowcroft J., "Quality of Service Routing for Supporting Multimedia Applications", *IEEE Journal on Selected Areas In Communications*, 14(7), p. 1228–1234, September 1996.

[WOL 95] Wolf L., Delgrossi L., Steinmetz R., Schaller S., Witting H., "Issues of Reserving Resources in Advance", *Proceedings of NOSSDAV*, Lecture Notes in Computer Science, p. 27–37, Durham, New Hampshire, April 1995, Springer.

[ZHA 97] Zhang Z., Sonchez C., Salkewicz B., Crawley E., "Quality of Service Extensions to OSPF or Quality of Service Path first Routing (QOSPF)", Internet draft (draft-zhang-qos-ospf-01.txt", work in progress, September 1997.

Chapter 3

Partitioning applications with agents

Oskari Koskimies and Kimmo Raatikainen
Department of Computer Science, University of Helsinki, Finland

1. Introduction

The environment of mobile computing is in many respects very different from the environment of the traditional distributed systems of today. Bandwidth, latency, delay, error rate, interference, interoperability, computing power, quality of display, and other non-functional parameters may change dramatically when a nomadic end-user moves from one location to another, or from one computing environment to another – for example from a wired LAN via a wireless LAN [4] (WLAN) to a GPRS [20] or UMTS [33] network. The variety of mobile workstations, handheld devices, and smart phones, which nomadic users use to access Internet services, increases at a growing rate. The CPU power, the quality of display, the amount of memory, software (e.g. operating system, applications), hardware configuration (e.g. printers, CDs), among other things ranges from a very low performance equipment (e.g. hand held organizer, PDA) up to very high performance laptop PCs. All these cause new demands for adaptability of data services. For example, palmtop PCs cannot properly display high quality images designed to be looked at on high resolution displays, and as nomadic users will be charged based on the amount of data transmitted over the GPRS network, they will have to pay for bits that are totally useless for them.

Software agent technology has gained a lot of interest in the recent years. It is widely regarded as a promising tool that may solve many current problems met in mobile distributed systems. However, agent technology has not yet been extensively studied in the context of nomadic users, which exhibits a unique problem space. The nomadic end-user would benefit from having the following functionality provided by the infrastructure: information about expected performance provided by agents, intelligent agents controlling the transfer

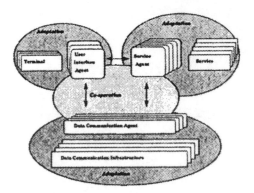

Figure 1. The adaptation triad

operations, a condition-based control policy, capability provided by intelligent agents to work in a disconnected mode, advanced error recovery methods, and adaptability.

The research project Monads [19] examines adaptation agents for nomadic users [26]. In the project we have designed a software architecture based on agents and we are currently implementing its prototypes. Our goal is not to develop a new agent system; instead, we are extending existing systems with mobility-oriented features. The Monads architecture is based on the Mowgli communications architecture [25] that takes care of data transmission issues in wireless environments. In addition, we have made use of existing solutions, such as FIPA specifications [14] and Java RMI [27], as far as possible. However, direct use was not sufficient but enhancements for wireless environments were necessary [3, 22].

By adaptability we primarily mean the ways in which services adapt themselves to properties of terminal equipment and to characteristics of communications. This involves both mobile and intelligent agents as well as learning and predicting temporary changes in the available Quality-of-Service (QoS) along the communications paths. The fundamental challenge in nomadic computing is dynamic adaptation in the triad service–terminal–connectivity (see Figure 1) according to preferences of the end-user.

The ability to automatically adjust to changes in the wireless environment in a transparent and integrated fashion is essential for nomadicity – nomadic end-users are usually professionals in other areas than computing. Furthermore, today's distributed systems are already very complex to use as a productive tool; thus, nomadic end-users need all the support, which an agent based distributed system could deliver. Adaptability to the changes in the environment of nomadic end-users is the key issue. Intelligent agents could play a significant role in implementing adaptability. One agent alone is not always able to make the decision how to adapt, and therefore adaptation is a cooperative effort carried out by several agents. Thus, there should be at least some level of cooperation between adapting agents.

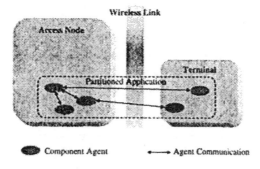

Figure 2. *Application partitioning*

Dynamic adaptation of a service to the properties of terminal equipment and available communication infrastructure is an attractive feature. We have previously explored predictive adaptation to available bandwidth with a Web browsing agent [26]: When the network connection is slow or predicted to become slow, the browser agent may automatically use different kinds of compression methods or even refuse to fetch certain objects.

We are now also examining adaptation to terminal equipment through *application partitioning*. With application partitioning we refer to the idea of dividing an application into component agents that communicate using e.g. FIPA ACL [16] messages. In this way, the application is running in a distributed fashion on both sides of the wireless link, as depicted in Figure 2.

Partitioning can be either *static*, *partially dynamic* or *fully dynamic*. Static partitioning, where component agent configuration is determined at compile time and cannot be changed, is of little interest to us. In *partially dynamic* partitioning, the location of a component agent is determined dynamically during application initialization, but cannot change during the application session. The most interesting is *fully dynamic* partitioning, which allows the component agents to be moved at any time during the application session. This is useful when bandwidth (or other dynamic resources such as memory) changes radically, and the change is expected to last for some time. For bandwidth, such a drastic change would typically be a vertical handover or a disconnection.

Partially dynamic partitioning is sufficient for adapting to different terminal types, but adapting to bandwidth changes (including disconnections) requires fully dynamic partitioning. Note that fully dynamic partitioning requires the component agents to be mobile, whereas partially dynamic partitioning does not.

2. Related work

In the 90s the notion of *nomadic computing* was introduced to launch information processing services accessible anytime and anywhere. According to Prof. Leonard Kleinrock [24], "Nomadic computing and communications will dramatically change the way people access information – and a paradigm shift in thinking about applications of the technologies involved. It exploits the advanced technologies of wireless, the Internet, global positioning systems, portable and distributed computing to provide anytime, anywhere access. It is beginning to happen, but it makes everything much harder for the vendors." He identified the following drivers:

Increased productivity: In many situations, the expenses associated with hiring, integrating, and developing staff account for much of the total cost entailed in deploying a given organizational capability.

"The personal touch": Another motivator is the social requirement for personal contact. The requirement for movement is rooted in the desire to meet, take the measure of, and enjoy the company of others.

Personal environment: The many parts of our lives – our family life, business life, personal social life, business social life, etc. – each frequently involves a different physical location. People move between these places as they move between the different aspects of their lives. Doing the appropriate thing in the appropriate place is key to the way many people organize their lives.

Interactivity: Much of the communication in intense human interactions is non-verbal and therefore requires face-to-face contact.

Setting: For various reasons, people frequently carry out business in locations other than in their own office.

Behind the notions of *ubiquitous computing* and *pervasive computing* there is almost the same problem space. The objective is to focus on services and to hide computing and communication. The ultimate goal is invisible computing and communication infrastructure on which services can be provided in various and dynamically changing environments.

In addition to seamless service provisioning, another significant trend is the requirement of ever-faster service development and deployment. The immediate implication has been the introduction of various service/application frameworks/platforms. Middleware is a widely used term to denote a set of generic services above the operating system. Although the term is popular, there is no consensus of a definition [8]. However, typical middleware services include directory, trading and brokerage services for discovery, transactions, persistent repositories, different transparencies such as location transparency and failure transparency.

The problem area of nomadic/pervasive/ubiquitous computing is vast and addressed by several research groups around the world. Below we briefly summarize most important ones that develop middleware infrastructure for future applications. The Endeavour Expedition at UC Berkeley [7] is a collection of projects that examines various aspects of ubiquitous computing. The goal is to enhance human understanding through the use of information technology, by making it dramatically more convenient for people to interact with information, devices, and other people. A revolutionary Information Utility, which is able to operate at planetary scale, will be developed. The underlying applications, which are used to validate the approach, are rapid decision making and learning. In addition, new methodologies will be developed for the construction and administration of systems of this unprecedented scale and complexity. In the middleware area the projects include:

- Ninja: Enabling Internet-scale Services from Arbitrarily Small Devices that develops a software infrastructure to support scalable, fault-tolerant and highly-available Internet-based applications [13].
- Iceberg: An Internet-core Network Architecture for Integrated Communications that is seeking to meet the challenge for the converged network of diverse access technologies with an open and composable service architecture founded on Internet-based standards for flow routing and agent deployment [9]
- OceanStore: An Architecture for Global-Scale Persistent Storage that is designed to span the globe and to provide continuous access to persistent information [10].
- Telegraph: An adaptive dataflow system that allows people and organizations to access, combine, analyze, and otherwise exploit data wherever it resides [11, 23].

In the MIT the corresponding project is called Oxygen [29]. The Oxygen project targets in the means of turning a dormant environment into an empowered one that allows the users to shift much of the burden of their tasks to the environment. The project is focusing on eight enabling technologies: new adaptive mobile devices, new embedded distributed computing devices, networking technology needed to support those devices, speech access technology, intelligent knowledge access technology, collaboration software, automation technology for everyday tasks, and adaptation methods. [6]

The project in the University of Washington at Seattle is called Portolano [31]. The project has three main areas of interest: infrastructure, distributed services, and user interfaces. An essential research area is data-centric routing that facilitates automatic data migration among applications. Context aware computing, which attempt to coalesce knowledge of the user's task, emotions, location, and attention, has been identified as an important aspect of user interfaces. Task-oriented applications encounter infrastructure challenges including resource discovery, data-centric networking, distributed computing, and intermittent connectivity. [1]

In the University of Illinois at Urbana-Champaign the research project in this area is 2K: A Component-Based, Network-Centric Operating System for the next Millennium [34]. The 2K is an open source, distributed adaptable operating system. The project integrates results from research on adaptable, distributed software systems, mobile agents and agile networks to produce an open systems software architecture for accommodating change. The architecture is realized in the 2K operating system that manages and allocates distributed resources to support a user in a distributed environment. The basis for the architecture is a service model in which the distributed system customizes itself in order to better fulfill the user and application requirements. The architecture encompasses a framework for architectural-awareness so that the architectural features and behavior of a technology are reified and encapsulated within software. Adaptive system software, which is aware of the architectural and behavioral aspects of a technology, specializes the use of these technologies to support applications forming the basis for adaptable and dynamic QoS, security, optimization, and self-configuration. [32]

The Mobile Computing Group at Stanford University (MosquitoNet [28] has developed the Mobile People Architecture (MPA) that addresses the challenge of finding people and communicating with them personally, as opposed to communicating merely with their possibly inaccessible machines. The main goal of the MPA is to put the person, not the device that the person uses, at the endpoints of a communication session. The architecture introduces the concept of routing between people by using the Personal Proxy. The proxy has a dual role: as a Tracking Agent, the proxy maintains the list of devices or applications through which a person is currently accessible; as a Dispatcher, the proxy directs communications and uses Application Drivers to massage communication bits into a format that the recipient can see immediately. It does all this while protecting the location privacy of the recipient from the message sender and allowing the easy integration of new application protocols. [12]

The PIMA project [30] at the IBM T.J. Watson Research Center has developed a new application model for pervasive computing. The model is based on the following three principles:

- A device is a portal into a space of applications and data, not a repository of custom software managed by the user.
- An application is a means to perform a task, not a piece of software written to exploit capabilities of a device.
- The computing environment is the information-enhanced physical surroundings of a user, not a virtual space to store and run software.

Based on those principles the following research challenges were identified:

- development of a programming model that identifies abstract interaction elements, specifying an abstract service description language, creating a task-based model for program structure, and creating a navigation model;
- building a development environment that supports the programming model above;
- developing specification languages for applications in terms of requirements, and for devices in terms of capabilities;
- developing mediating algorithms to negotiate a match between application requirements and device capabilities;
- run-time detection of changes in available resources and redistribution of computation;
- handling temporary disconnections; and
- enhancing current techniques of failure detection and recovery.

3. The need for terminal adaptation

The QoS of a wireless link can vary wildly, due to interference, network load and vertical handovers. In some cases, like vertical handover, these changes can be quite drastic. In order to provide smooth operation, an application needs to adapt to changes in QoS. Traditionally, the adaptation is done by compressing and reducing content that is transferred to the terminal.

However, adapting to just QoS is not enough. Unlike the users of fixed networks, who almost all have fairly similar, workstation-class terminals, the terminals of nomadic users vary from smartphones to powerful laptops. Even the same user is likely to use different classes of terminals, e.g. a laptop on business trips and a PDA or smartphone during his free time. However, currently the nomadic user often has to use a different application for the same purpose on different classes of terminals. For example, the calendar software on a laptop is likely to be very sophisticated and offer an advanced graphical user interface, whereas the software on a smartphone would probably offer only minimal functionality. Even though the different applications may be able to exchange information, the situation is still undesirable for two reasons:

1. An application is limited to the terminal class it was designed for, and
2. Users have to learn a different application for each terminal class.

The application can, of course, be implemented separately for each terminal class, but this wastes implementation effort, and a version for smaller terminals may have drastically reduced functionality. What is required is an application that can adapt to different terminal classes without sacrificing functionality.[1]

---

1. However, *usability* may suffer. No amount of adaptation can make the keyboard or screen of a smartphone match that of a laptop.

4. Adaptation by partitioning

We are examining adaptation to terminal equipment and QoS variation by partitioning an application into components. Because adaptation of the overall application requires that the individual components can cope with varying component configurations, the components must both be themselves adaptive, and also have a degree of autonomy. Thus, it makes obvious sense to model the components as agents. To give an example: an email application can work with different types of terminals by partitioning the application in different ways, depending on terminal and wireless link characteristics. Assume the email application consists of four agents:

1. *User interface agent*: Either a standard FIPA UDMA agent [17] or a dedicated UI agent that has an advanced graphical user interface and supports voice input.
2. *Core email agent*: Handles basic email processing.
3. *Email filtering agent*: Groups and prioritizes messages, and handles any automated email processing.
4. *Email compression agent*: Compresses messages prior to transferring them to the terminal (this includes lossy methods such as leaving out attachments).

With low-performance equipment (a high-end mobile phone, for example), only the user interface (a standard FIPA UDMA agent) of the application runs on the terminal. If the terminal has more capabilities (like a PDA), also the core email agent can run on the terminal, with the email filtering and compression agents running on the network side. Finally, with a laptop terminal, all agents except the compression agent can be run on the terminal.

Partitioning can be used for adapting to available bandwidth as well. For example, if bandwidth is very low, it is better to run the filtering agent on the network side, since it may be able to handle some emails automatically, and prioritizing means the user will get the more important messages first. On the other hand, if bandwidth is high, it is better to run the filtering agent on the terminal, since that allows it to more easily communicate with the user about filtering decisions, enabling more fine-grained filtering control. By using mobile agents, this kind of adaptation can be dynamic, with agents moving to optimal locations while the application is running. We call this *repartitioning*.

Another facet of partitioning is the possibility to use different agents for different terminals, or not to use a particular agent at all. For example, with a high-performance laptop, a large, dedicated email UI agent can be used instead of a small, generic FIPA UDMA agent. Or, when bandwidth is high, the use of an email compression agent becomes unnecessary.

4.1 Assumptions

The design of our partitioning system is based on a few assumptions:

Applications are specially designed We assume that the application is specially designed to support partitioning, for the following reasons:

1. An application that wasn't designed to be partitioned is unlikely to be easily divided into separate independent components.
2. Partitioning, and especially repartitioning, requires the component agents to be designed flexible so that reconfiguration is possible. It is unrealistic to expect this from applications in general, although applications built by component composition (from "off-the-shelf" component agents) in the future might have the required flexibility.
3. Practical agent applications are still scarce, so one can assume that the most important agent applications are still to be built, and their design is thus still open.

Application metadata is available Since applications are already assumed to be specially designed for partitioning, it is not unreasonable to assume that some meta-data (such as agent resource requirements) has been made available as well.

Repartitioning is heavy and uncommon The actual time taken by any single repartitioning operation depends on available bandwidth and the size of the agents involved, but we assume that repartitioning is a heavy operation, due to both its transactional nature and the need to transfer agent state. Thus, repartitioning would be worthwhile only when drastic changes in circumstances occur (e.g. vertical handover).

Note that the initialization of a partitioned application is a much lighter operation, but may still require application code to be transferred over the wireless link.

4.2 How to partition an application

Our project has previously produced a system for predicting near-future QoS fluctuations [26]. We utilize these predictions, along with profiles, to make partitioning decisions.

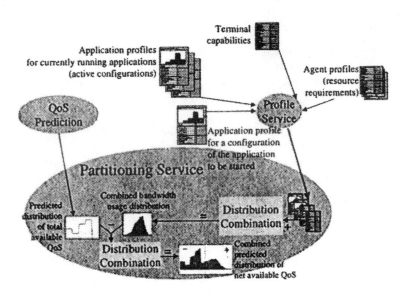

Figure 3. Partitioning decision

4.2.1 Profiles

Three types of profiles are used:

Application profiles The Application profile lists the component agents in the application, and a set of possible configurations for them. For each configuration, agent locations and a communication profile are given, as well as a *utility* value that represent how "good" (in terms of usability) that configuration is for the user. If the application is willing to share some agents with other applications, that information would be here also.

    Note that alternative (such as different GUI agents) and optional (such as compression agents) agents can be represented by configurations where all agents are not included.

Agent profiles Agent profiles contain resource requirements for the agent, as well as startup cost and repartitioning (movement) costs. Additionally, if the agent cannot move and/or be replaced, that information is stored here.

Terminal profiles The capabilities of terminals, such as memory and screen size, are listed in terminal profiles.

4.2.2 Bandwidth distributions

Figure 3 illustrates how a partitioning decision is made. When an application is started, the partitioning service first loads the application profile. The list of agents is then used to get the resource requirement profiles for the agents. These requirements are compared against the terminal capability profile to get a list of possible configurations for this terminal, and the communication profiles for those configurations.

In simple terms, for each possible application configuration, its bandwidth use (from the communication profile) is added to the bandwidth use of currently running applications, to get combined bandwidth use. This is then deducted from the predicted total available bandwidth, to get the predicted net available bandwidth. This is the prediction of the remaining bandwidth after the application is started.

The reality is more complex, however: A prediction of available bandwidth is not a single value but a distribution that gives the probability for getting a specific bandwidth. Likewise, the bandwidth use of an application varies according to a probability distribution. Thus, the result of the above calculation is also a distribution.

If the resulting distribution gives a high probability for having a negative net (remaining) bandwidth, that means that the configuration that is being examined cannot be run with the available bandwidth. Otherwise, the resulting distribution is given a *utility score* that is based on how much bandwidth was taken by it, and the utility value of that configuration for the user (available from the application profile). The score is penalized if the configuration exhausts available resources on the terminal. The configuration with the best final score is then selected.

4.2.3 Simulation

Unfortunately, adding and subtracting distributions is not trivial. Also, if a simple histogram-type of distribution is used to represent the bandwidth use of an application, we do not take into account that the bandwidth use often varies with time. For example, there may be an initial surge of bandwidth use during application initialization, to be followed by more moderate use. If this kind of variation is taken into account, the distribution calculations quickly become unwieldly.[2] As an alternative, we propose to use simulation.

If simulation is used, then application profiles contain serialized application simulator objects instead of bandwidth use distributions. The simulator objects are used by the partitioning service to simulate a run of the application. By including the simulator objects of currently running applications in the simulation, together with a prediction of future QoS, we get an estimate of remaining bandwidth. By running the simulation several times, we can increase the reliability of the estimate.

The advantage with simulation is that calculations are simple (unless very complex models are used in the simulator objects), and it allows more accurate

---

2. We are using a relatively simple method of calculation, however. Research into more mathematically sophisticated methods might yield improvements in efficiency.

modeling of application behaviour. Furthermore, the amount of processing devoted to evaluating a configuration can be adjusted by varying the number of simulation runs and the length of the simulation time unit. Also, the utility of a configuration can be determined dynamically by the simulation object. The disadvantage is that the statistical error in the results is probably higher, since the number of repeated simulations that can be made is limited.

A simulation object contains at least a generic model (common for all users) for the bandwidth use of the application. In addition, the application itself can update the simulation object with user-specific models. As a simple but effective way of doing this, the application can collect traces of bandwidth use, and add them to the simulation object. Others, e.g. terminal-class-specific models, could also be used.

Additionally, simulation objects keep track of the QoS the simulated application receives during a simulation run, and provide a utility value to the partitioning service at the end of the simulation run. The utility value describes how well the application was able to function in the simulation.

The modified decision-making process is depicted in Figure 4. First, the partitioning service collects from the profiles all the simulation objects for the active configurations of currently running applications. Then it tries the possible configurations for the new application one by one, simulating each in turn.

When simulating a configuration, the partitioning service first selects a simulation time unit based on the processing power available, and decides how long one configuration can be simulated. At the start of the simulation, each simulation object is initialized and given the relevant terminal profiles. Then, for each simulation round:

1. Each simulation object is given the time period the round corresponds to.
2. The simulation object returns the amount of data it wants to send that round. If data transfer priorities are used by the application, then those must be returned as well.
3. Each simulation object is then given the bandwidth available to the application that round. Any scheduling policy in use (e.g. priority-based) has to be taken into account at this point.
4. Finally, the simulation object then gives its current estimate of the overall utility of the application at this point. Depending on user preferences (minimum utility values), a low utility value may cause the simulation to be terminated and restarted with another configuration.

At the end of the simulation, the final overall utility is marked for the configuration. If there is still time, the simulation is redone. The utility for the configuration is then the average of the utilities from the different runs. If a significant fraction (depending on user preferences) of the simulation runs results in the

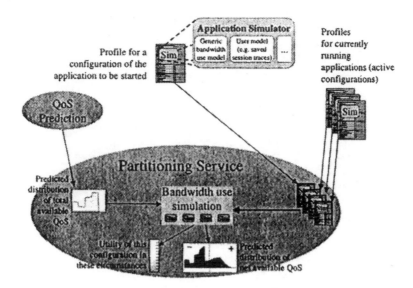

Figure 4. Partitioning decision (modified)

application becoming unusable (utility below user-defined acceptance level), the configuration is rejected. When all configurations have been simulated, the one with the best utility is chosen.

4.3 Starting the application

The application startup sequence is shown in Figure 5. The original startup request (1) is directed to the partitioning service. First, the partitioning service selects a unique ID for the application session. Then it queries the yellow pages (e.g. a FIPA DF [15]) in the terminal and on the network side (2) to see if some agents are already running and previously registered (2b), and then issues create-agent requests for those agents that are not running (or cannot be shared) (3).

To start an agent, the partitioning service may use a Factory service that advertises itself via the Yellow Pages, or it can contact the agent platform directly (e.g. a FIPA AMS [15]). Either way, a new agent is created (4). If agent profiles contain a startup cost, the fact that an agent is running can also be taken into account when selecting configurations. When an agent has successfully initialized itself, it sends a message (5) to the partitioning service to confirm that it is ready.

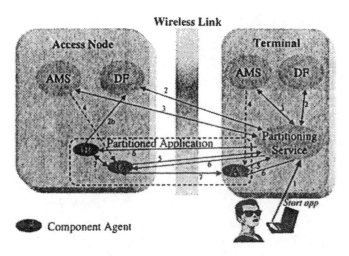

Figure 5. Application startup

Finally, the application session ID is sent to all participating agents, together with information about their partners (6), binding them together to form the application. The application session ID is especially important for shared agents, who use it to manage their state information. Application communication can now start (7). In fully dynamic partitioning, this phase can be redone during repartitioning to remove the need for rerouting messages of moved agents.

4.4 Repartitioning

The weakness of partially dynamic partitioning is that while the partitioning decision depends on both terminal capabilities and the QoS of the wireless connection, only terminal capabilities are relatively static. QoS may change drastically during the lifetime of an application session, making a previously made partitioning decision invalid. Fully dynamic partitioning becomes necessary: The application must be *repartitioned* by moving its component agents to new locations. Thus, mobile agents are required.

In fully dynamic partitioning, the decision making process is rerun when QoS is predicted to change drastically. Frequent reruns of the process can be avoided by ignoring smaller changes, or by only considering vertical handoffs and disconnections.

If a rerun of the decision making process shows that another configuration is superior to the current one, the difference of the utility scores of the configurations (the profit), multiplied with the time the new conditions are predicted to last, is compared to a penalty calculated from the repartitioning costs of the agents. If the profit is greater, repartitioning is initialized.

The actual repartitioning process, shown in Figure 6, is somewhat like a two-phase commit (see e.g. [18]). First, the partitioning service sends each component agent a partitioning request that contains the application session ID and the agent's orders (to stay, move, be replaced or shut down), and asks if the agent is able to participate. The agent checks if it can free itself of any application state that cannot be carried over the repartitioning process, and sends a 'yes' answer if it can. Once the answer has been sent, the agent enters a state where it waits for the partitioning process to complete, refusing requests not related to the partitioning process. Note that even though an agent answers 'no', it should not continue normal operation, since other agents in the application may not be ready to continue yet.

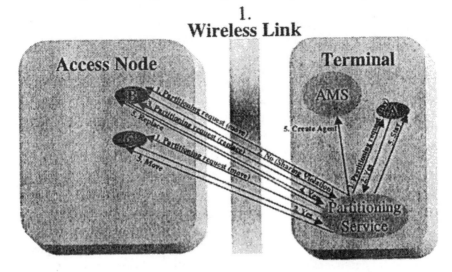

Figure 6. Repartitioning

Error handling follows the normal two-phase commit procedure, with one exception: an agent may be unable to move because it is shared by other applications. In that case, it sends back a no answer, and additionally indicates that the refusal is due to a sharing violation. The partitioning service may then immediately (without aborting) resend the partitioning request to the agent, but with replacement orders instead of movement orders. Since the additional message exchanges takes time, service-type shared agents should preferably be marked as non-movable in the agent profile.

If the partitioning service gets a yes answer from all agents, it then sends each agent a message which tells them what to do. Once an agent has completed the order, it sends back a reply to the partitioning service to signal that the order has been successfully executed. If the agent platform does not support adequate message rerouting after agent movement, the reply may contain the new messaging address of the agent.

If the new configuration requires agents that were not present in the old one, these are created the same way as in application startup, described in the previous section. If an agent in the old configuration is not present in the new one, its order will be to shut down, and it will send the reply just before shutting down.

Finally, once all agents have reported in, the partitioning service sends each agent a 'continue' message to tell them that the repartitioning has completed, and informs them of changed messaging addresses and new, removed or replaced agents. The agents can now continue from where they left off. Note that in some cases this is not straightforward; for example, a simple user interface agent may have been replaced with a more complex one, or some agents may not be present in the new configuration. However, it is up to the agents themselves to adapt to the new situation. Note that this implies that component agents really must be *agents*; mere objects would not have the required adaptability.

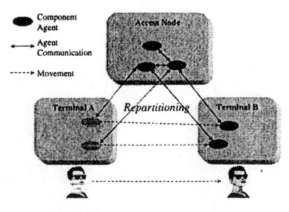

Figure 7. *Personal mobility*

4.5 Repartitioning and personal mobility

As shown by Figure 7, repartitioning can also be used for personal mobility. This requires only minor extensions to the system described above. Firstly, the partitioning service on terminal A must fetch the terminal profile for terminal B, and use it instead of the terminal A profile. Secondly, all agents that the repartitioning process would have placed on terminal A, are issued orders to move to terminal B. Agents that are specific to a terminal type may be replaced by others, but that is a normal part of the partitioning process. Finally, the partitioning service on terminal A must pass on the responsibility for the application to the partitioning service on terminal B.

Table 1. *Example agents*

Agent	Description	State Information	Footprint
UDMA	Generic user interface agent	10 kB	100 kB [4]
DGUI	Dedicated UI	10 kB	1 MB
Core	Core Email Agent	100 kB	1 MB
Filter	Email Filtering Agent	20 kB	200 kB
Compr	Email Compression Agent	–	200 kB

4.6 An example scenario

To give an idea of the communication costs involved in partitioning, we have built a prototype of the partitioning system, using the JADE 1.4 [2, 5] agent platform.[3] It was tested with a test application consisting of four agents, mimicking the email application example. Each agent had only the functionality necessary for the partitioning process, and some dummy state data.

The agents, their state data and memory requirements (as contained by the profiles) are described in Table 1. The test scenario is outlined in Figure 8:

1. The user is in her office, using a desktop computer with a 100 Mbps LAN connection. She starts reading her mail. The partitioning system chooses a configuration with a dedicated UI agent[4] and without an email compression agent, where everything is running on the terminal, and starts the email application.

---

3. The reason for using Jade is that it is also otherwise used by our project. However, JADE's implementation of agent mobility is not optimized, so using it also has the benefit of getting conservative results.
4. The generic user interface agent can be shared by other applications.

Figure 8. Example scenario

2. The user reads her mail. After a while, she comes across one that requires a reply, and switches to voice input, dictating the reply message.

3. After dictating about half of the reply, she notices it is time to go home, and switches to her palmtop, which has a 2 Mbps WLAN connection in the office, and a 28.8 kbps GSM High Speed Data connection [21] while outside.[5] The partitioning system moves the email application to the palmtop, using now a configuration with a generic user interface agent. The partial reply is saved to the state of the core email agent, and moved with it.

4. The user finishes writing the reply while waiting for the elevator. Because a generic UI agent is used, the GUI of the email application is now less polished, and no longer accepts voice input, but the user can continue to write the reply using normal text input.

5. Based on the time, and the fact that the user left the office, the QoS prediction system [26] gives a high probability that bandwidth will soon drop dramatically. The partitioning system recalculates the optimal configuration. The resulting configuration has an email compression agent and the email filter agent on the network side, and the other agents in the terminal.

6. The user continues to read her email using this configuration while sitting on a train. Although the user interface of the email application is more bare (generic UI), attached images lose some detail or are omitted (compression), and messages from mailing lists are ignored (filtering), she is still using the

---

5. All these connection types are already commercially available.

same application, and even the same application session, as when she started reading her email.

The scenario was tested with our prototype. The test was run using a Pentium III Linux workstation as the access node, a Pentium II Linux workstation as the desktop terminal and a Pentium Linux laptop as the palmtop computer. A real WLAN was used, but the GSM data link was simulated with software. The results are given in Table 2. Note that the times are median values from five test runs.

As can be seen, the communication delays are quite acceptable for this scenario. On the last partitioning, however, that is due to the successful QoS prediction that allowed repartitioning to be initiated while bandwidth was still high. If the repartitioning had been done *after* the drop in QoS, when bandwidth was down to 28.8 kbps, the repartitioning would have taken 14 seconds – still acceptable, but a very noticeable delay. Note that application classes were already present on access node and terminals.

Table 2. Test results

Event	Bandwidth	State data	Time
Application start (1)	100 Mbps	–	0.7 s
Terminal change repartitioning (3)	2 Mbps	120 kB	4.3 s
QoS change repartitioning (5)	2 Mbps	20 kB	2.2 s

5. Conclusions and future work

We have shown how partitioning can be used to adapt an application to varying QoS and terminal capabilities. Partitioning can be used for implementing personal mobility, as well. Our first tests indicate that the solution is feasible as far as communication costs are concerned.

Our next step will be to fully implement the partitioning system and use it in the Monads QoS prediction system, so that the prediction system can be optimally configured for different terminal types. The problem of how to minimize state loss during repartitioning is also worthy of attention. Finally, partitioning-related messaging must be optimized to reduce the partitioning overhead.

REFERENCES

[1] M. ESLER ET AL. Next Century Challenges: Data-Centric Networking for Invisible Computing. In *Proceedings of MobiCom99*, p. 256–262, Seattle, Washington, August 1999.

[2]   F. BELLIFEMINE, G. RIMASSA, AND A. POGGI. JADE – A FIPA-compliant Agent
      Framework. In *Proceedings of the Fourth International Conference and
      Exhibition on the Practical Application of Intelligent Agents and Multi-Agents
      (PAAM 1999)*, April 1999.

[3]   S. CAMPADELLO, H. HELIN, O. KOSKIMIES, AND K. RAATIKAINEN. Performance Enhancing
      Proxies for Java2 RMI over Slow Wireless Links. In *Proceedings of the Second
      International Conference and Exhibition on the Practical Application of Java (PA
      Java 2000)*, April 2000.

[4]   B. P. CROW, I. WIDJAJA, J. G. KIM, AND P. T. SAKAI. IEEE 802.11 Wireless Local Area
      Networks. *IEEE Comm. Mag.*, pp. 116–126, September 1997.

[5]   Jade Web Site. Available electronically from http://sharon.cselt.it/projects/jade/

[6]   M. DERTOUZOS. The Oxygen Project. *Scientific American*, 281(2): 52–63, August
      1999.

[7]   Endeavour Expedition: Charting the Fluid Information Utility. Available
      electronically from http://endeavour.cs.berkely.edu/

[8]   B. AITKEN et al. Network Policy and Services: A Report of a Workshop on
      Middleware. *IETF RFC 2768*, February 2000.

[9]   H. J. WANG et al. An Internet-core Network Architecture for Integrated
      Communications. *IEEE Personal Communications*, August 2000.

[10] J. KUBIATOWICZ et al. OceanStore: An Architecture for Global-Scale Persistent
      Storage. In *Proceedings of the Ninth International Conference on Architectural
      Support for Programming Languages and Operating Systems (ASPLOS 2000)*,
      November 2000.

[11] J. M. HELLERSTEIN et al. Adaptive Query Processing: Technology in Evolution. *IEEE
      Data Engineering Bulletin*, 2000.

[12] P. MANIATIS et al. The Mobile People Architecture. *ACM Mobile Computing and
      Communications Review*, July 1999.

[13] STEVEN D. GRIBBLE et al. The Ninja Architecture for Robust Internet-Scale Systems
      and Services. *Computer Networks (Special Issue on Pervasive Computing)*. To
      appear.

[14] Foundation for Intelligent Physical Agents. FIPA Web Site. Available electronically
      from http://www.fipa.org/

[15] Foundation for Intelligent Physical Agents. FIPA 97 Specification Part 1: Agent
      Management, October 1997. Available electronically from http://www.fipa.org/

[16] Foundation for Intelligent Physical Agents. FIPA 97 Specification Part 2: Agent
      Communication Language, November 1997. Available electronically from
      http://www.fipa.org/

[17] Foundation for Intelligent Physical Agents. FIPA 98 Specification: Human-Agent Interaction, 1998. Available electronically from http://www.fipa.org/

[18] A. GOSCINSKI. *Distributed Operating Systems: The Logical Design*, chapter 5.4.8, pp. 203–204. Addison-Wesley, 1991.

[19] Monads Research Group. Monads Web Site. Available electronically from http://www.cs.helsinki.fi/research/monads/

[20] GSM Technical Specification, GSM 02.60. GPRS Service Description, Stage 1, 1998. Version 6.1.0.

[21] GSM Technical Specification, GSM 03.34. High Speed Circuit Switched Data (HSCSD), Stage 2, May 1999. Version 5.2.0.

[22] H. HELIN, H. LAAMANEN, AND K. RAATIKAINEN. Mobile Agent Communication in Wireless Networks. In *Proceedings of the European Wireless '99 Conference*, October 1999.

[23] J. M. HELLERSTEIN AND R. AVNUR. Eddies: Continuously Adaptive Query Processing. In *Proceedings of the ACM SIGMOD 2000 Conference*, 2000.

[24] L. KLEINROCK. Nomadicity: Anytime, Anywhere in a Disconnected World. *Mobile Networks and Applications*, 1(4): 351–375, January 1997.

[25] M. KOJO, K. RAATIKAINEN, M. LILJEBERG, J. KIISKINEN, AND T. ALANKO. An Efficient Transport Service for Slow Wireless Telephone Links. *IEEE Journal on Selected Areas in Communications*, 15(7): 1337–1348, September 1997.

[26] M. MÄKELÄ, O. KOSKIMIES, P. MISIKANGAS, AND K. RAATIKAINEN. Adaptability for Seamless Roaming Using Software Agents. In *XIII International Symposium on Services and Local Access (ISSLS2000)*, Stockholm, Sweden, June 2000.

[27] Sun Microsystems. Java Remote Method Invocation – Distributed Computing for Java. White Paper, 1998.

[28] MosquitoNet: The Mobile Computing Group at Stanford University. Available electronically from http://mosquitonet.stanford.edu/

[29] MIT Project Oxygen. Available electronically from http://www.oxygen.lcs.mit.edu/

[30] The PIMA Project: Platform-Independent Model for Applications. Available electronically from http://www.research.ibm.com/PIMA/

[31] Portolano: An Expedition into Invisible Computing. Available electronically from http://portolano.cs.washington.edu/

[32] M. ROMAN AND R. H. CAMPBELL. Gaia: Enabling Active Spaces. In *Proceedings of the 9th ACM SIGOPS European Workshop*, Kolding, Denmark, September 2000.

[33] Third Generation Partnership Project Web Site. Available electronically from http://www.3gpp.org/

[34] 2K: An Operation System for the Next Millennium. Available electronically from http://choices.cs.uiuc.edu/2k

Chapter 4

Mobile agents for adaptive mobile applications

Thomas Kunz, Salim Omar and Xinan Zhou

Systems and Computer Engineering, Carleton University, Ottawa, Ontario, Canada

1. Introduction

Mobile computing is characterized by many constraints: small, slow, battery-powered portable devices, variable and low-bandwidth communication links. Together, they complicate the design of mobile information systems and require the rethinking of traditional approaches to information access and application design. Finding approaches to reduce power consumption and to improve application performance is a vital and interesting challenge. Many ideas have been developed to address this problem, ranging from hardware to software level approaches.

Designing applications that adapt to the challenges posed by the wireless environment is a hot research area. One group of approaches concentrates on mobile applications that adapt to the scarce and varying wireless link bandwidth by filtering and compressing the data stream between a client application on a portable device and a server executing on a stationary host. Some [BOL 98] enhance the server to generate a data stream that is suited to the currently available bandwidth. Others [ANG 98, FOX 98] extend the client-server structure to a client-proxy-server structure, where a proxy executes in the wireless access network, close to the portable unit. This proxy transparently filters and compresses the data stream originating from the server to suit the current wireless bandwidth.

A second set of approaches provides general solutions that do not change the data stream, focusing on improving TCP throughput [BAL 95]. They usually treat IP packets as opaque, i.e., they neither require knowledge of, nor do they exploit information about, the data stream. While this addresses issues such as high link error rates and spurious disconnections, it does not address the low bandwidth offered by most wireless technologies, nor does it address the problem of limited resources at the portable device.

We propose a third, complementary approach, focusing not on the data stream but on the computational effort required at the portable device. Mobile applications, especially ones that are used for intensive computation and communication (such as next-generation multi-medial PCS and UMTS applications), can be divided dynamically between the wired network and the portable device according to the mobile environment and to the availability of the resources on the device, the wireless link, and the access network. The access network supports the mobile application by providing proxy servers that can execute parts of the application [WAN 00]. This may potentially increase the performance of applications and reduce the power consumption of portable devices since offloading computation to the proxies in the wired network will reduce their CPU cycles and memory requirements [OMA 99].

This paper discusses our mobile code toolkit and demonstrates the feasibility of this idea by reporting on our experience with a resource-intensive mobile application, an MP3 player. The results show that both increased application performance and reductions in power consumption are possible under certain conditions by encapsulating the resource-intensive decoding in a mobile agent and migrating it to the less constrained access network.

The paper is organized as follows. Section 2 reviews toolkits to support adaptive mobile applications based on mobile agents/code. Section 3 presents our mobile code toolkit. Performance improvements and power reductions achievable under certain environment conditions for our MP3 player are the topic of Section 4. Section 5 discusses the scalability of our approach and Section 6 summarizes our main contributions and highlights future work.

2. Related work

Mobile applications need to be capable of responding to time-varying QoS conditions. In the following sub-sections, we briefly describe popular tools and middleware that support adaptive mobile applications and contrast their approach to our work.

Comma [KID 98] provides a simple and powerful way for application developers to access the information required to easily incorporate adaptive behavior into their application. It provides easy-to-use methods to access this information, a wide variety of operators and ranges available to provide the application the information it needs when it needs it, a small library to link with to minimize the overhead placed on the client and to minimize the amount of data that needs to be transferred between the clients and the servers.

The Rover toolkit [JOS 95] offers applications a distributed-object system based on the client-server architecture. The Rover toolkit provides mobile communication support based on re-locatable dynamic objects (RDOs). A re-locatable dynamic object is an object with a well-defined interface that can be dynamically loaded into a client computer from a server computer, or vice versa, to reduce client/server communication requirements.

Sumatra [RAN 96] is an extension of the Java programming environment. Policy decisions concerning when, where and what to move are left to the application. The high degree of application control allows programmers to explore different policy alternatives for resource monitoring and for adapting to variations in resources. Sumatra provides a resource-monitoring interface, which can be used by applications to register monitoring requests and to determine current values of specific resources. When an application makes a monitoring request, Sumatra forwards the request to the local resource monitor.

Mobiware [ANG 98] provides a set of open programmable CORBA interfaces and objects that abstract and represent network devices and resources, providing a toolkit for programmable signaling, adaptation management and wireless transport services. Mobile applications specify a utility function that maps the range of observed quality to bandwidth. The observed quality index refers to the level of satisfaction perceived by an application at any moment. The adaptation policy captures the adaptive nature of mobile applications in terms of a set of adaptation policies. These policies allow the application to control how it moves along its utility curve as resource availability varies.

In general, it is left to the application to decide how to react to environment changes. This argues for exporting the network state as well as available resources of the portable device to the mobile applications to be designed to be adaptive. On the other hand, the automation of adaptation to the resources was not explored. There are a number of similarities between our work and the work in Sumatra. Both Sumatra and our work use extended Java Virtual Machines for portability and the ease of use of the language especially for implementing object mobility toolkits. The main difference is that, in our work, adaptation to the change in the resources and environment is partially left to the toolkit. We try first to use available remote resources to achieve the same task; otherwise, as a last resort, we let the application do the adaptation.

3. Mobile code toolkit

The central concept of our framework is the proxy server. An application can use the resources of the proxy server to increase the performance and decrease the power consumption by executing selected objects on the proxy server. To enable this approach, we need to identify computationally intense, closely-coupled groups of objects and encapsulate them in a mobile agent. This agent may execute locally or be shipped to the access network, depending on the wireless link conditions and the available resources at both the portable device and the proxy server.

3.1 Toolkit design

Our goal is to extend the Java Virtual Machine with a toolkit that facilitates the mobility of objects between portable device and proxy server in a dynamic manner, transparent to the application designers and users. This toolkit is designed to work on PDAs with a limited amount of memory. Other existing ORBs and object mobility toolkits do not support these platforms or they have too big a memory footprint.

The toolkit has a set of APIs, which provides the required functionality for moving objects dynamically. One instance of the toolkit executes on both the portable device and the proxy. It contains the following major modules:

– Monitor: monitors and delivers the state of the portable device and the proxy server as events to the Object Server. Changes in the bandwidth or changes in the power status are examples of the events that this unit exports.
– Code Storage: storage of the validated classes files (bytecode) at the portable device.
– Object References and Profiling: representation of the application's objects along with the profiling information about these objects.
– Object Server: core of the toolkit. It runs a thread that listens continuously to all the commands from a remote object server. Commands can be related to moving objects or related to the remote invocation of a method on a remote object.
– Remote Method Invocation Protocol: marshal and un-marshal a method's parameters.
– Dynamic Decision: analyzes the profiling information of application's objects. It resides only at the proxy server.
– Communication Control Layer: simulates wireless links in terms of low bandwidth. We introduce a controllable amount of delay between data packets, which allows us to control the throughput dynamically at run time for testing purposes.

In our toolkit, every movable application object is associated with a proxy object that has the same interface. Other objects will not reference application objects directly, but they reference them through their proxies. Assume a simple application with two objects, A and B. Initially, four objects reside at the portable device: A, B, A's proxy, and B's proxy. B references A, so in our toolkit B will contain a reference to the proxy of A. Moving B from the portable device to the proxy server will not require moving A to the proxy server as well. However, at the proxy server, a proxy of A must be created to forward the calls to A at the portable device. Also, a new proxy of B will be instantiated at the proxy server to allow local objects there to reference B.

Should the toolkit migrate A to the proxy server, changing the reference to A in B is not required since B references only the proxy of A (which will remain behind at the portable device). Any calls from B to A will be forwarded remotely through the proxy at the portable device.

3.2 Distributed garbage collection

Every proxy object created in the toolkit is assigned a local and a remote reference counter. These counters are updated whenever a proxy object is referenced locally or remotely and are used to determine when the garbage collector can claim the proxy and its associated object. Whenever a proxy object is not being referenced remotely and locally, it will be finalized and garbage collected. If the associated object of this proxy is local, then the associated object will be finalized and claimed again by the garbage collector as well. If the associated object is remote, then the proxy object will inform the remote object server to decrement the remote reference counter for the associated object at the remote site, which in turn garbage collects the object if there are no further references locally or remotely to the associated object.

3.3 Toolkit performance

To obtain a first impression of our toolkit performance, we performed a few simple tests, comparing various aspects with Voyager [OBJ 99], a popular mobile code toolkit. The following table compares the measured overhead of the toolkit against Voyager. The measurements were taken under Windows NT on a Pentium II 350 MHz PC.

Table 1. Overhead comparison

	Voyager	**Our toolkit**
Footprint	2620 KB	204 KB
Moving object	142 ms	80 ms
Calling a method	23 ms	110 ms

Based on these results, we are confident that our toolkit can be used for small portable devices such as Palms and PDAs. The memory requirement of our toolkit is small compared to Voyager (which supports many additional features such as CORBA compliance), allowing it to be embedded in these small devices. The cost of migrating the objects compared to Voyager is lower as well; however, we still need to improve the remote method invocation protocol.

3.4 Automatic agent identification

In the final version of the toolkit, we also expect to provide support for the automatic identification of suitable mobile agents. In essence, suitable mobile agents are sets of application objects that consume lots of CPU time and communicate heavily. Currently, we are profiling applications with a Java profiler, derive information about objects, their CPU consumption, and their interaction patterns, and use an external clustering algorithm to determine a suitable set of application objects to group as agent. This external definition of an agent is then used by the mobile code toolkit. For our MP3 player, Table 2 lists the instances (objects) that exist during the decoding of MP3 frames and some information about their CPU time consumption. This profiling information was measured on a 350 MHz Pentium II PC.

Table 2. MP3 player profiling information

Object Name	Data Size (bytes)	# of Instances	Code Size (bytes)	Calls/Frame	Avg. CPU Time/Instance
Table43	28	1	107344	620	0.00794
Bit_Reserve	16666	1	1430	3355	0.50058
SBI	223	6	2905	97	0.01926
gr_info_s	376	4	5195	7184	0.06222
temporaire2	376	2	1409	1687	0.00445
Temporaire	593	2	1819	52	0.22175
Header	765	1	9245	11	0.00267
III_side_info_t	959	1	2022	35	0.16715
Ibitstream	1972	1	5301	449	0.29177
huffcodetab	2526	35	45493	693	0.11331
SynthesisFilter	4414	2	18724	4824	0.44725
LayerIII_Decoder	25114	1	47146	160	1.07052

The profiler also collects information about the number of method invocations and the number and size of parameters exchanged. Most objects communicate only relatively infrequently, but some communicate frequently, with lots of data being passed. Unfortunately, the heavily communicating objects are not necessarily the most CPU-intensive objects. The clustering algorithm therefore starts with the computationally heavy objects and adds additional objects to the group to reduce "coupling". Some objects, such as those implementing the GUI component or the one interacting with the local sound card, cannot be migrated to the proxy server. In our application environment, "coupling" measures the amount of inter-object communications between the mobile agent and the remaining application objects. Should the toolkit decide to migrate the agent to the proxy server, such inter-object communication will result in remote method invocations, and we aim to minimize the traffic over the slow wireless link. In the future, we hope to integrate profiling into our toolkit to allow us to determine appropriate agent candidates dynamically at runtime. This also enables us to track changes in the application behavior and to react to them appropriately.

4. Case study

We implemented an MP3 player in Java to demonstrate the feasibility of our general approach. This application requires a powerful CPU to decode the sound file due to the complexity of its encoder/decoder algorithm, which makes it an ideal candidate to demonstrate the need for fast static hosts, i.e. proxy servers, to support the relative resource-constrained portable devices.

We executed the MP3 player under various emulated environment conditions and observed application performance and power consumption on the portable device. Based on the observed environment, our runtime system instantiates some objects on the proxy server, others are created on the portable device. We studied in particular the following two parameters:

– Available Bandwidth.
– Relative CPU speeds (Portable Device: Proxy Server).

To observe the importance of the first parameter, the bandwidth available, we choose 19.2 Kb/sec to represent CDPD [CDP 95], a typical wide-area cellular data service. For high bandwidths we choose 1000 Kb/sec to represent the bandwidth that can be obtained from Wireless Ethernet cards such as WaveLan [LUC 99].

To observe the importance of the second parameter, the relative CPU speed, we fixed the bandwidth to 1000 Kb/sec, so it does not represent a scarce resource. A Windows CE PDA and a laptop were used as portable devices. The PDA contains a RISC processor at 75 MHz and the laptop runs a Pentium processor at 133 MHz. The proxy server runs on a 350 MHz Pentium II PC. The performance of Java applications depends primarily on the performance of the JVM. Both laptop and the proxy server run high performance JVMs. The JVM on the PDA, on the other hand, is very slow, so the relative CPU speed degrades considerably. We measured the relative CPU speed between the PDA, laptop and the proxy and found it to be 1:116 and 1:4, respectively.

The experiments are based on decoding MP3 coded audio frames, with the assumption that output is mono, with a sampling rate of 11025 Hz, and 16 bits per sample (which will impact the network traffic when decoding is done remotely and is based on the achievable quality of sound-cards in low-cost handheld devices).

The detailed experiments and results are discussed in depth in [OMA 00]. These results demonstrate that available bandwidth is an important factor. If bandwidth is the bottleneck in the system, neither reduction in power consumption nor increases in MP3 player performance can be obtained, no matter what the relative CPU speed. However, if bandwidth is not the bottleneck, then the relative CPU speed becomes a decisive factor in increasing the performance and decreasing the power consumption of the portable device. It is possible to save power and increase performance of the MP3 player if the entire decoder is to be executed remotely and the PDA only works as sound player. These improvements can be very dramatic: the application will execute up to 20 times faster if decoding is done at the proxy, consuming only 5% of the power it would take to decode the MP3 file on the PDA. On the other hand, using the laptop as a portable device, an MP3 player that decodes locally always performs better than decoding at the proxy, even though the available bandwidth is sufficient to handle the decoded sound and the computational power is quite high at the proxy server.

Overall, the results show that it is not always beneficial to ship mobile agents to more powerful proxies to gain performance and/or decrease power consumption. The benefits depend on available bandwidth and relative CPU speed. We also expect them to depend on the graph topology and the data traffic volume between application objects. For our sample application, an MP3 player, migrating the decoding component to a more powerful proxy leads to a considerable decrease in power consumption as well as an increase in the performance when executing on a Windows CE PDA; however, it is not worth migrating the MP3 decoder to a proxy server when using a laptop as portable device.

5. Scalability

To support our approach, proxy servers need to be deployed throughout the access network, which could be a large provincial or national cellular network. On the one hand, one could envision an architecture where each wireless cell provides dedicated proxy servers, resulting in relatively little concurrent use of an individual server but inducing a high handover overhead and costs. At the other extreme, we could provide only one or very few proxy servers that support applications in many different wireless cells, reducing the handover overhead but requiring more powerful servers. With potentially multiple thousands of users executing resource-intensive next-generation mobile applications, the scalability of our approach becomes extremely important. To explore this issue, we started to develop performance prediction models based on Layered Queuing Networks (LQNs).

LQNs study the performance of distributed systems that have hardware and software [FRA 98, ROL 95]. A task is a basic unit in LQN. A task represents the software in execution. An entry represents a service provided by a task. If a task can provide multiple services, then the task has multiple entries. Entries in different tasks communicate with each other through requesting and providing services. Client tasks make requests to proxies; these in turn invoke services provided by the application server task. Requests are either synchronous or asynchronous.

Each task has at least one entry. The entries in a task have execution demands respectively, and may also interact with entries in other tasks by calling the entries in those tasks. The client tasks will not receive requests from other tasks. They are called *reference tasks*. For reference tasks, usually there is a think time that is denoted by Z, which implies the pause time between two consecutive operations. Execution demands of entries are characterized in two phases. Phase 1 is the service demand between reception and response (for synchronous requests), phase 2 describes the service demands after the response (for synchronous requests) or the total service demand for asynchronous requests. The LQN analytical tool describes the system by the average behaviour of the entries and solves the performance model by approximate MVA calculations. To study the scalability of our system, we developed a four layer LQN, extracted data from traces collected from an operational WAP-based system [9], and studied the impact of introducing proxy servers. In modelling terms, the introduction of a proxy results in less execution demand on the portable devices and more execution demand on the proxy servers. Since we assume the proxy servers to be more powerful, the increase in load is only fraction of the load decrease on the portable device.

The complete model is shown in Figure 1. A parallelogram represents a task entry. Cascaded parallelograms indicate an entry of multiple tasks. The task name is written near the parallelogram. [Z] in the client task entry models the client think time. [0, tc] in the client task entry represents the execution demands of the client task entry between requests. The pair of brackets inside the non-referential task entries has the same meaning as the one in the client task entry. The notation under the pair of brackets is the entry name. The ellipse represents CPU processors. The arrow segment connects the calling entry and the called entry. The straight segment connects the task and the CPU on which the task runs. The pair of circular brackets beside the arrow line contains the number of calls from the calling entry to the called entry. 'sh' denotes synchronous calls and 'ay' denotes asynchronous calls. Client tasks make (indirectly) requests to an application server and wait for the responses. This server answers some of the requests directly and passes some to other servers on the Internet. Generally, passing a request to another server takes less time than answering one directly. The application server task has four entries.

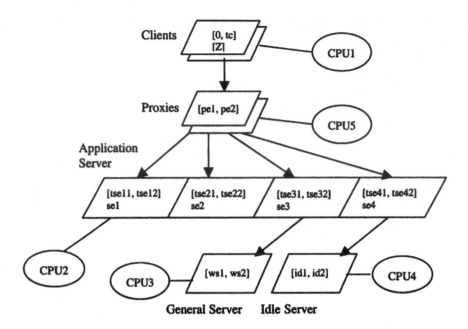

Figure 1. Layered queuing model

The first entry se1 processes the synchronous requests from client tasks and responds to the clients directly. The second entry se2 is responsible for asynchronous requests from client tasks. The third entry se3 passes synchronous requests from the clients to other servers. The General Server task is used to represent additional servers on the Internet since it is impossible to get information for all the Internet servers and model them individually. The fourth entry se4 is used to represent the idle time between consecutive sessions with the help of an imaginary Idle Server task and CPU4.

We studied the capacity of the system under various conditions and the effect of transferring execution load from handsets to proxies. The capacity of the system indicates the maximum number of users that the system can serve at the same time without being saturated. Proxies, Application Server, General Server and Idle Server are multithreaded, and can provide as many threads as needed. We define 0.7 as the threshold utilization of CPU2, beyond which we consider that the system is saturated. 'MC' is the maximum number of clients that the system can sustain in this case. We studied the effect of increasing the percentage of requests serviced directly by the application server and executing a higher proportion of the client tasks at the proxy servers.

We traced the WAP-based application for several months; some detailed data is reported in [KUN 00]. The total average number of synchronous requests (the sum of sh1 and sh2) per user session is 11.2, i.e., sh1 + sh2 = 11.2. sh1 = 0 represents the case when all the requests are passed on to other servers. sh1 = 11.2 represents the case when all the requests are directly processed by the Application server. We show the capacity implications of various splits between sh1 and sh2 in Table 3. To derive the base system capacity, we assumed that the proxy layer is essentially non-existent: requests from clients and replies from the Application server are forwarded immediately and no processing happens at the proxy.

The base capacity of the system decreases with the increase of sh1. That is, the more requests the Application server processes directly, the smaller the system capacity becomes. For the period we traced the application, the average maximum numbers of concurrent sessions for each month are indeed below 8, but during the peak hours of some days, the maximum number of concurrent sessions is bigger than 10. This period of potential capacity overload is very short, however, usually less than one minute.

Table 3. Basic system capacity

sh1	0	1.2	2.2	3.2	4.2	5.2	6.2	7.2	8.2	9.2	10.2	11.2
MC	50	36	29	24	20	17	15	13	11	10	9	8

We also investigated the effect of load migration from handsets to the proxies. We assume that the proxy CPU is 25 times faster than the handset CPU. The load migration from handsets to the proxies reduces the service demand at the clients' side. Assuming that each user has access to a dedicated proxy server, this is equivalent, in modeling terms, to replacing a slow user with a faster, more demanding user (more requests per unit time), reducing the overall system capacity. However, since proxy servers are shared between multiple users, this may not necessarily be the case. We varied the service demand on the portable device (tc) from 6 to 0, in steps of 1, with a corresponding (smaller) increase in service demand (pe1) at the proxy. The performance prediction results are shown in Tables 4 and 5.

Table 4. Capacity vs. load migration, all client requests processed by application server

Tc	6	5	4	3	2	1	0
Pe1	0	0.04	0.08	0.12	0.16	0.2	0.24
MC	8	7	7	7	7	6	6

Table 5. *Capacity vs. load migration, all client requests forwarded to external servers*

tc	6	5	4	3	2	1	0
pel	0	0.04	0.08	0.12	0.16	0.2	0.24
MC	50	46	43	42	41	40	39

We can see that, all else being equal, the capacity decreases with increasing migration of computational load from portable devices to the proxies. This is consistent with other result reported in [ZHO 00] that show that the system can serve more slow users than fast ones.

6. Conclusions and future work

Finding approaches to reduce power consumption and to improve application performance is a vital and interesting problem for mobile applications executing on resource-constrained portable devices. We suggested a new approach in which part of an application will be encapsulated in a mobile agent and potentially shipped for execution to proxy servers, according to the portable device and fixed host's available resources and wireless network state. To support our approach, we designed and developed a mobile object toolkit, based on Java. With this toolkit we combine JVMs on both the proxy server and the portable device into one virtual machine from the application point of view. The results showed that it is possible to simultaneously improve application performance and reduce power consumption by migrating the entire MP3 decoder to the proxy server in the case of a slow portable device and sufficient wireless bandwidth.

To study the scalability of our approach, we developed a Layered Queuing Model, derived trace data from a live system, and studied the maximum number of concurrent users that can be supported. Even in a system with many potential users, our traces reveal that only a relatively small number of users are concurrently accessing the application server. In these cases, the introduction of proxy servers does not overly reduce system capacity. Other studies have shown that for the system studied, the application server is more likely to become a bottleneck, rather than the proxy server. While these findings are application-specific, they are encouraging. Contrary to our initial suspicions, we will probably not need proxy servers in each cell to support the user population. In all likelihood, a few centrally placed proxy servers can support potentially many users.

A number of issues need to be addressed in future work, some of which is currently under way. We are working on improving the mobile object toolkit. The main improvement to our toolkit optimizes the RMI protocol. Another improvement deals with proxy objects. To support location-transparent invocation of methods, each object is associated with one or more proxy objects. Currently, we manually write these proxy objects; however, we plan to develop tools to automate this process (similar to Voyager) and integrate it with the toolkit. Another addition to the toolkit would be to integrate profiling to enable dynamic agent identification, based on computational demands and inter-object communication patterns. Finally, we are currently porting the toolkit to Palm OS, which requires fundamental changes to the object migration component due to the limitations of the available Java Virtual Machines (no reflection mechanism, no object serialization).

A second issue is to explore scenarios where either the application behavior or the execution environment changes drastically while the application executes. Intuitively, we would expect the runtime system to rebalance the application accordingly. However, migrating agents/objects at runtime is not cheap. So we need to explore how to balance the resulting overhead with the anticipated performance gains and power reductions, in particular in execution environments that change rapidly. Also, while our results reported here show that there is no trade-off between power reduction and performance improvement, previous work [OMA 99] indicates that there may well be such trade-offs for other applications. In these cases, we need to identify how to balance conflicting objectives. One possible solution could be to allow the mobile user to select preferences that prioritize objectives.

A third area is the refinement of the performance prediction model. The current model is essentially based on a multi-tier client-server architecture. In cases where the mobile agent acts like a server (decode the next MP3 frame and return the sampled sound), this accurately reflects the application structure. In more general cases, though, objects at the proxy server side will request services from objects that remain at the portable device. Another refinement of the model would include the wireless link as additional "service" layer between client tasks and proxies, to capture contention for that shared and scarce resource.

Yet another area currently under investigation is the extension of our work to support truly mobile users. The work reported here focuses on the migration of application components between a portable device and a more powerful proxy server. Should more than one proxy server be required to support a given user population, we need to concern ourselves with handoffs between proxies. As described in [WAN 00], such handoffs can occur for a number of reasons: proxy servers can become overloaded, different proxy servers may be closer to a roaming user and therefore be better suited to provide better service, etc. Our toolkit needs to be expanded to include agent migration between proxies and we need to identify and test appropriate migration strategies.

A final area of possible future work is the interaction between application-aware and application-transparent adaptation. Our MP3 player does not react to changes in bandwidth, for example by reducing sampling size or audio quality. In our experiments, we fixed the output playing rate and the sampling size. Further study is required to show how application adaptation policies affect and interact with the automated adaptation by our toolkit.

Acknowledgements

This work was support by the National Research and Engineering Council, Canada (NSERC) and a research grant by Bell Mobility Cellular, a Canadian wireless service provider.

REFERENCES

[ANG 98] ANGIN O. et al., "The Mobiware Toolkit: Programmable Support for Adaptive Mobile Networking", *IEEE Personal Communications*, 5(4), August 1998, p. 32–43.

[BAL 95] BALAKRISHNAN H. et al., "Improving TCP/IP Performance over Wireless Networks", *Proceedings of the 1st Int. Conf. on Mobile Computing and Communications*, Berkeley, USA, November 1995, p. 2–11.

[BOL 98] BOLLIGER J., GROSS T., "A Framework-based Approach to the Development of Network-aware Applications", *IEEE Trans. on Software Eng.*, 24(5), May 1998, p. 376–390.

[CDP 95] CDPD Consortium, *Cellular Digital Packet Data System Specification,* Release 1.1, January 19, 1995 (CD-ROM).

[FRA 98] FRANKS G., WOODSIDE M., "Performance of Multi-level Client-server Systems with Parallel Service Operations", *Proceedings of the 1st Int. Workshop on Software and Performance (WOSP98)*, Santa Fe, October 1998, p. 120–130.

[FOX 98] FOX A. et al., "Adapting to Network and Client Variation using Infrastructure Proxies: Lessons and Perspectives", *IEEE Personal Comm.*, 5(4), August 1998, p. 10–19.

[JOS 95] JOSEPH A. D. et al., "Rover: a Toolkit for Mobile Information Access", *ACM Operating Systems Review*, 29(5), December 1995, p. 156–171.

[KID 98] KIDSTON D. et al., "Comma, a Communication Manager for Mobile Applications". *Proceedings of the 10th Int. Conf. on Wireless Communications*, Calgary, Alberta, Canada, July 1998, p. 103–116.

[KUN 00] KUNZ T. et al., "WAP Traffic: Description and Comparison to WWW Traffic", *Proceedings of the 3rd ACM Int. Workshop on Modeling, Analysis and Simulation of Wireless and Mobile Systems (MSWiM 2000)*, Boston, USA, August 2000.

[LUC 99] Lucent, *WaveLAN Wireless Computing*, http://www.wavelan.com/

[OBJ 99] ObjectSpace, *Voyager 2.0.0 User Guide*, http://www.objectspace.com/Voyager/

[RAN 96] RANGANATHAN L. et al., *Network-aware Mobile Programs*, Technical Report CS-TR-3659, Dept. of Computer Science, University of Maryland, College Park, MD 20740, June 1996.

[OMA 99] OMAR S., KUNZ T., "Reducing Power Consumption and Increasing Application Performance for PDAs through Mobile Code", *Proceedings 1999 Int. Conf. on Parallel and Distributed Processing Techniques and Applications*, Vol. II, Las Vegas, Nevada, USA, June 1999, p. 1005–1011.

[OMA 00] OMAR S., *A Mobile Code Toolkit for Adaptive Mobile Applications*, April 2000, Thesis (M.C.S.), Carleton University, School of Computer Science.

[ROL 95] ROLIA J. A., SEVCIK K. C., "The Method of Layers", *IEEE Transactions on Software Engineering*, 21(8), August 1995, p. 689–700.

[WAN 00] WANG J., KUNZ T., "A Proxy Server Infrastructure for Adaptive Mobile Applications", *Proceedings of the 18th IASTED Int. Conf. on Applied Informatics*, Innsbruck, Austria, February 2000, p. 561–567.

[ZHO 00] ZHOU, X., *Cellular Data Traffic: Analysis, Modeling, and Prediction*, Master's Thesis, School of Computer Science, Carleton University, July 2000.

Chapter 5

Active networks: architecture and service distribution

Nicolas Rouhana
University Saint-Joseph, Beirut, Lebanon

Eric Horlait
University Pierre et Marie Curie, Paris, France

1. Introduction

The past ten years has witnessed the rapid development of computer networks, accompanied by an increased demand for new value-added services to meet the highly varying requirements of end-users. For example, new alternative levels of service are currently needed at various points of the Internet to provide a better than traditional "best-effort" service, such as admission control, DiffServ and IntServ/RSVP mechanisms, MPLS technology, QoS Routing, as shown in Figure 1 [1].

The need for these services has resulted in more and more functionality having to be deployed inside the network, which lead to engineer programmable networking infrastructures that offered open and extensible programming interfaces providing abstraction between hardware and software. The IN concept [2] was a solution that emerged from the telecommunications sector and defined open interfaces to the switching control plane, thus easing the deployment of third party novel control software and services. The *Opensig* [3] community is also based on open-interfaces and virtual node abstractions, and regroup projects such as the *IEEE Project 1520* [4] and Columbia University *xbind* [5] project allows the programmability of management and control planes in diverse networks.

In IP networks, the DARPA community introduced *Active Networks* in which service construction is based on code mobility through "active" packets that

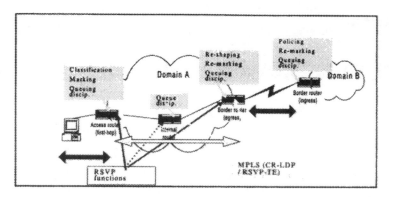

Figure 1. *Example of new services in today's Internet*

contain not only data, but also code, and "active" nodes that perform customized computations depending on packets contents (Figure 2). The packet-switching paradigm hence evolved to become "store, *compute* and forward"; i.e., the traditional router or switch responsible for "passive" header processing and routing, now present a transient *execution environment* that evaluates and executes code in the "active" packets. Network behavior and service construction can thus be controlled dynamically and at run-time on a per-packet, per-user, or other basis, rather than a programmable control plane.

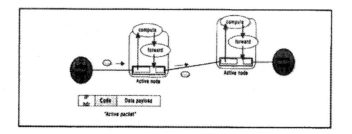

Figure 2. *Active networks = Σ active nodes + active packets*

A key enabler for active networks was the advance in technologies such as ultra-fast hardware, safe and efficient code languages and mobility schemes. This created a research environment in which diverse active networks architectures [6] were experimented with different execution environments and programming languages that could be applied in "real networks". In parallel, efforts have been underway at DARPA to define and standardize a new active network architecture [7] aiming to supplant the current Internet architecture. The next section details the functional elements of the DARPA architecture, while Section 3 will give examples of existing architectures and how they relate to the proposed framework. We present future works and conclusions in Section 4.

2. Architecture of an active node

In the draft framework, the functionality of an active network node is divided between Execution Environments (EEs), a Node Operating Systems (NodeOS) and Active Applications (AAs). Figure 3 shows the modular architecture of the active node "reference model". By analogy with a traditional OS, the NodeOS is responsible for allocating and controlling the node's physical resources (e.g. bandwidth, memory, CPU cycles), and each Execution Environment interprets incoming active packets that contain the necessary code constituting the active application. The detailed role of each component is discussed in the following paragraphs.

2.1 Node Operating System

The NodeOS constitutes a layer of abstraction between the Execution Environments and the underlying physical resources of the node (i.e., CPU cycles, memory, transmission bandwidth, etc.). A NodeOS can be layered on top of any

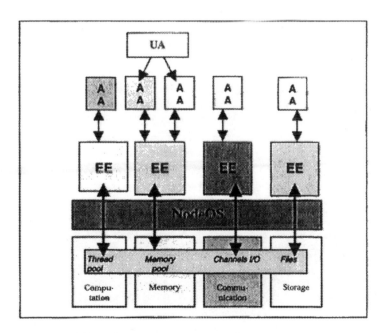

Figure 3*. Generic components of an active node*

traditional OS that provides the access to the node's low-level resources; the existence of the NodeOS is therefore necessary to provide a base level of functionality common to every active node independent of the underlying OS and hardware used.

The NodeOS also adds mechanisms to the underlying OS to provide novel and interesting services to support active networking:

– supports multiple competing execution environments in the node and provides basic safety between them;
– provides configurable task-scheduling and level of concurrency between the execution environments;
– provides the execution environments with fine-grained control over scheduling or over accountability, meaning that the NodeOS must be able to identify individual packet flows, and limit how much resources a particular packet flow or EE is able to consume.

Traditional operating systems, of course, provide more services, such as protection between programs running in different address spaces, or some priority-based scheduling policy favoring interactive programs over long-running batch-jobs. The NodeOS should be able to work equally well with a host OS that does or does not provide additional services; i.e., Execution Environments can make use of special capabilities in the underlying OS and hardware (e.g. multi-processing, real-time, etc.), but it should also be possible for them to exist on just a set of minimal services.

[8] defines an abstract NodeOS-EE interface providing access to the node's resources through a system-calls interface. This interface consists of five primary resource abstractions: *thread pools, memory pools, channels, files* and *flows*. The first four abstractions are mapped by the NodeOS to actual physical resources, respectively being computation, memory, communication and storage. The fifth abstraction is used to aggregate control and scheduling of the other four abstractions. Typically, each execution environment creates, runs and destroys a *flow*, which consists of a set of input and output channels through which messages are received and sent, a memory pool, and a thread pool. Active packets arrive on an input channel, are processed by their corresponding execution environment using threads and memory allocated to the flow, and then transmitted on an output channel. The abstract communication *channels* actually consist of physical transmission links (e.g. Ethernet, ATM) plus higher-level protocol stacks (e.g. TCP, IP, link protocols). The NodeOS classifies incoming packets based on specified criteria on the packets header defined by the EEs or the AAs, and forwards the packets to the appropriate channel. Packets can also follow a *cut-through* path and directly forwarded to an output channel if no treatment is required at this particular node.

2.2 Active transport

Some means is required to allow an active node receiving a packet to uniquely and quickly determine the environment in which it is intended to be evaluated. We already stated that the NodeOS can use a certain criteria specified by each EE (e.g. based on specific fields in the packets headers) to classify the packets, or can make use of a separate protocol at the NodeOS level that demultiplexes directly the execution environment.

Early active networks architectures, prior to the node model with NodeOS, used the "overlay" method, whereby networks were emulated by connecting the active nodes and their execution environments with UDP channels, (e.g. ANTS [9]), providing the paths for the active packets. Another scheme is to use the ActiveIP [10] option, which defined a new option to the IP datagram header, allowing embedding of a program fragment in an IP datagram that gets evaluated and executed in "active" routers, while "legacy" routers just forward the datagram.

Both these methods relied on IP-based networks. A more general approach is to use a transport protocol independent of the technology used (i.e. the protocol can be placed over Ethernet, IP or UDP), and that ensures interoperability with existing networks. Two protocols have been proposed in that direction: the *Active Network Encapsulation Protocol (ANEP)* [11] and the *Simple Active Packet Format (SAPF)* [12].

2.2.1 Active Network Encapsulation Protocol

ANEP is a protocol for encapsulation of active network frames, designed to provide the capability to identify the different evaluation environments, and to allow minimal, default processing of packets for which the intended evaluation environment is unavailable. Furthermore, information that does not fit conceptually or pragmatically in the encapsulated program, such as security headers, can be placed in the header. The ANEP header includes a Type Identifier field that indicates the EEs at the node. The proper authority for assigning public Type ID values to interested parties is the Active Networks Assigned Numbers Authority (ANANA).

ANEP also provides a vehicle for other communications with the NodeOS, including:

- Error-handling instructions. When a packet fails to reach the desired execution environment at a node (perhaps because the node does not support it, or resources were not available), the ANEP header lets the user instruct the Node OS as to the expected response, for example: drop the packet silently, try to forward it, or send an error message.
- Security vouchers. The ANEP header is a natural location for a node-to-node credential. It is impractical for every node in the network to retrieve or store the

information (e.g. public key certificate) required to authenticate every packet passing through it. However, it may be practical for a packet to be authenticated just once (based on its originator) when it enters the network, and thereafter be vouched-for node-to-node using keys shared between neighboring nodes.

2.2.2 Simple Active Packet Format

The SAPF protocol is based on an identifier carried by the packet that directly points to the associated execution code (like a function handler) in the node. The motivation behind SAPF is to enable a generic NodeOS to remove considerable packet handling overhead from each active execution environment: currently, EEs parse the ANEP format themselves whilst SAPF moves all (destination specific) packet fields to the payload field except the one that allows for demultiplexing the packet to the targeted EE.

The active network cloud consists of nodes that exclusively exchange SAPF packets containing SAPF selectors. Well-known selectors will be statically assigned, but the majority of the field values is dynamically assigned with the help of active packets that are exchanged via the few statically assigned selector values. An intermediate active node assigns a locally unique selector to this handler and communicates it back to the upstream node. The upstream node forwards all subsequent packets belonging to the same IP flow to this "SAPF channel". In order to interwork with IP networks, the ingress active node handles an IP flow using active networking mechanisms instead of simple IP forwarding. In this case, the node would create a handler in one of the active network EEs that transforms each IP packet into an active packet or does some other packet processing.

2.3 Execution Environments, Active Applications and User Applications

The *Execution Environments* evaluate and execute the code in the packets. They provide the *network APIs* through which users can create end-to-end high-level services. The execution environment can be, for example, a Java Virtual Machine or any language interpreter, and the framework provides support for multiple and diverse execution environments. One reason for this multiplicity is that the limitation to one standard byte code and execution semantics would yield too much functionality and complexity for a common element, and would probably not reach a consensus in the active network community. The interface between the execution environment and the active application is by definition EE-specific, and thus reflects its "programming paradigm". Some execution environments may offer services that are accessible via a narrow, language-independent interface (as with sockets); others may require that the active application code be written in a particular language (e.g. Java). The intention is that this interface should be more or less independent of the underlying active network infrastructure.

Active Applications are the programs run by the execution environments and define the actual end-to-end service associated with the packets. They implement the customized services for *End-User Applications* (e.g. congestion control, reliable multicast), using the programming interface supplied by the execution environment.

The execution environments extend the role of the NodeOS "upwards" into user-space. They distribute their available resources among the applications they launch, control the execution of applications and prevent boundary and resource violations between them, and can allow the applications to deliberately share information. The NodeOS and the EE must cooperate to dispatch active packets received to the appropriate active applications and to send their packets into the network.

Execution environments can provide for simultaneous access to more than one service. As a somewhat artificial example, consider an active application that provides mobility and another that provides reliable group communication. Ideally, an end-user should be able to invoke both services simultaneously, which is a very challenging problem for today's implementations.

Service deployment and distribution

A fundamental requirement of active networking is that the Active Application code be dynamically loadable over the network. Schemes for distribution and downloading of code are developed so that not all nodes need to have the code that they may use. The code can be carried within protocol data packets, called the *integrated* approach, or it can be resident in the node and loaded out-of-band from the protocol data, called the *discrete* approach. The former approach offers maximum flexibility in support of service creation, i.e. at packet transport granularity, but with the cost of adding more complexity in the programming model in providing safe transient execution environments.

The *discrete* approach is definitely preferable when programs are relatively large compared to the packets, and maintains a modularization between user data and program, which may be useful for network management tasks. Details of how the "code" constituting the active application is loaded into the relevant nodes of the network are determined by the Execution Environment. Out of-band loading may happen during a separate signaling phase or on-demand, upon packet arrival; it may occur automatically (e.g. when instructions carried in a packet invoke a method not present at the node, but known to exist elsewhere) or under explicit control. Also, users can send a program "off-line" to a particular node, and when a packet arrives at that node, the corresponding program is selected using header information and then executed at the node. When a new version of the program is necessary, or if a different type of processing is required, the user can send the new program to the node to replace the old one.

3. Related architectures

In this section, we provide a survey on some active networks implementing some or all of the functions and mechanisms related to the previous section, namely NodeOS, Execution Environments, and Active Applications.

3.1 Odyssey architecture

The *Odyssey* active networking environment is an implementation of the DARPA active networking architecture. It consists of two major components: the *Bowman* [14] Node Operating System and the CANEs [13] execution environment (Figure 4). Odyssey assumes a host operating system (e.g. Solaris, Linux) over which a node operating system is built. Execution environments and generic services that are likely to be useful across environments (e.g. routing) are built using the abstractions provided by the node operating system; user code and particular protocols are run as part of an execution environment.

Bowman is a NodeOS design implementing a subset of DARPA NodeOS interface. It is developed in user-space level and exports to EEs an API based on three resource abstractions: channels, a-flows and state stores, with functions for manipulating these abstractions such as create, run, delete, etc. Bowman channel is a communication end point of abstract links; it consists of a set of protocols (TCP, UDP, IP, etc) and optionally local and remote addresses supporting abstract topologies that constitutes user-defined connectivity between Bowman nodes over physical topologies, thus enabling network-wide abstractions used to implement virtual networks. A-flows abstractions in Bowman represent channel-specific computation, and consist of at least one thread, processing contexts, user states and time-out routines. Execution environments may create a-flows per-user or

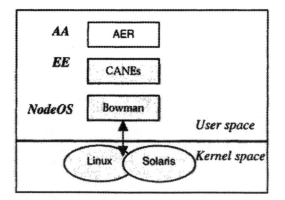

Figure 4. *Components of an Odyssey node*

may just create one a-flow hence the EE itself becomes just one user to Bowman (equivalent to a "process" in Unix). Flow-specific processing is based on a high-performance and configurable packet classifier that identifies the set of channels on which the received packet should be processed. The state-store provides a mechanism for a-flows to store and retrieve state for data-sharing between them without sharing program variables.

The CANEs execution environment executes as an a-flow within Bowman in Odyssey and provides a composition framework for active services based on a foundation of one or more generic behaviors (the underlying program) that can be customized by injecting code (the injected program) to run in specific points called slots. Composition is achieved in two steps; in the first the user selects an "underlying program" that is executed on behalf of the user. Forwarding functions are typical examples of underlying programs. Depending on the user, and the network provider, there may be several forwarding functions available at each node, each of which would define a specific set of processing slots. Depending upon their needs, users may choose different underlying programs. The users then select/provide a set of "injected programs", which correspond to individual code modules, that can be used to customize the underlying program on a per-user basis. In effect, slots identify specific points in the underlying program where the injected code may be executed. Slots are also the mechanism by which the underlying program asserts properties that hold at the node. As in an event-driven framework, injected code is "bound" to specific processing slots, and becomes eligible for execution when appropriate slots are "raised". Both the underlying and injected programs are demand loaded, dynamically. CANEs also provides mechanisms for variable sharing between programs exporting and sharing slots.

Active Error Recovery (AER) [15] is an attempt to develop a reliable multicast framework using active processing. AER is an active application that makes use of a repair server residing within the network to cache packets, respond to retransmission requests, suppress redundant NAKs from receivers, and detect gaps in sequence numbers indicating lost packets. AER also includes protocols to calculate round trip times and dynamically select a worst receiver to handle sliding-window-based flow control. AER is implemented in CANEs using two underlying programs, a multicast forwarding engine for sending data, NAKs and source path messages, and the generalized forwarding function for calculating the round trip time and monitoring congestion status.

The Bowman channel, a-flow, and state-store abstractions are too elementary to be useful for most users. Thus, Bowman provides an extension mechanism that is analogous to loadable modules in traditional operating systems. Using extensions, the Bowman interface can be extended to provide support for additional abstractions such as queues, routing tables, user protocols and services. Bowman does not enforce safety constraints on loaded code at run time. During the signaling phase in Bowman, the code-fetching mechanism must decide, based on credentials provided by the user and the node security policy, whether it trusts the requested code; if not, the request is denied.

3.2 Active Network Node

The software architecture of ANN [16] supports two EEs: ANTS [9] (for network management and experimental prototyping of network protocols), and Distributed Code Caching for Active Networks (DAN [17]), running on top of a NodeOS, as shown in Figure 5. ANN has been developed as part of the Router Plugins project [18], which is an architecture allowing code modules (compiled object code in 'C'), called plugins, to be dynamically added and configured at run time within active routers.

In these active routers, the regular IP forwarding loop is extended to look for special headers between the IP header and the UDP or TCP header, which reference plugins. If such a header is found, the packet is passed to the referenced plugin prior to being forwarded. If the referenced plugin is not present on the system, it is downloaded from a code server over the network and automatically installed in the kernel by the DAN on-demand downloading of plugins scheme. The Active Applications are built by deploying new plugins, like the WaveVideo Plugin and Application that uses an active router plugin to provide congestion control by doing intelligent scaling for a wavelet video transmission [19].

ANN uses as a base for NodeOS a modified NetBSD Unix kernel implementing IPv4/IPv6 and QoS functionality in a modular fashion,

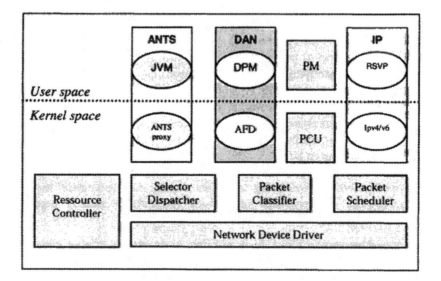

Figure 5. The ANN software architecture

incorporating user-processing at specific points (gates) on the IP forwarding path. Components of the NodeOS permitting flow-specific processing include: a fast packet filter that classifies incoming packets to specific flows and identifies the function that is bound to a packet at each processing gate, a DRR Packet Scheduler, a Selector Dispatcher (to demultiplex packets containing selectors instead of IP headers), and a Resource Controller that keeps track of the CPU cycles and memory consumed by plugin instances and distribute fair CPU time sharing among active functions. The packet processing path in Router Plugins is integrated with the flow-specific processing, Router Plugins do not incur thread-switching and queuing overheads.

Architecturally, Router Plugins merge Node OS and EE functionality and thus Router Plugins do not export a node OS interface over which different execution environments can be built. The major feature that is missing from Router Plugins is the general set of communication abstractions – channels, abstract links and topologies reflecting the IP-based focus of the Router Plugins project. The proposed framework introduces an execution environment which is a userspace software component for code downloading, called Plugin Management (PM). The plugin management includes the following sub-components: an Active Module Loader (AML) which loads the active modules authenticated and digitally signed by their developers from well known code servers using a lightweight network protocol (e.g. UDP/IP); a Policy Controller which maintains a table of policy rules set up by an administrator, e.g. restrict the set of supported modules; a Security Gateway which allows/denies active modules based their origin and developer by analyzing their digital signatures/authentication information; a Module Database Controller which efficiently administers the local database of active modules, and an Active Function Dispatcher (AFD) which identifies references to active modules in data packets and passes these packets to their corresponding function implementations.

3.3 Active Reservation Protocol

The ARP project [20] is exploring the use of portable and dynamically-extensible protocol code for network control protocols, especially for signaling protocols. That is, ARP is developing a facility for dynamic composition of new or modified service features with an existing signaling service. In terms of the active network architecture under development by the DARPA research community, the ARP project is building an Execution Environments appropriate for active signaling, as well as significant Active Applications to execute in that environment (Figure 6).

At present, the architecture runs on a traditional OS and the Execution Environment is called the *Active Signaling Protocol (ASP)*. The ASP is essentially a user-mode operating system, providing services and resources to the Active Applications executing within it. These services are defined by a

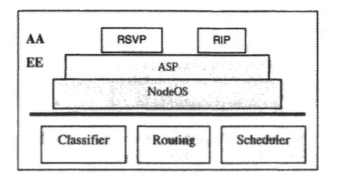

Figure 6. The ARP architecture

Protocol Programming Interface (PPI) that the EE presents to its applications. With ARP, an initial or "base version" version of the code that implements a particular signaling algorithm can be modified by one or more "functional extensions" (FEs). The base version plus the set of FEs in use for a particular signaling activity define a "version" of the signaling code. An appropriate version may be selected for each distinguishable signaling activity; for example, for RSVP, an appropriate version may be used for each RSVP session, with the base version as default. New versions may be installed and controlled by network management, which would exert administrative control over the mapping of sessions onto versions by instalation of appropriate classifiers. Ultimately, this would allow individual customers of the signaling function to be able to develop and dynamically install their own versions, which requires solving the protection and security problems inherent in allowing arbitrary code fragments in the node.

Security and isolation are important issues in the design of the ASP EE. Like the active applications, the ASP EE is written in Java providing class-based protection, and isolation, and the standard Java "sand-boxing" severely restricts the mischief that an application can cause. The architecture does not use the capsule model, but instead fetches AA code out of band from the flow of active packets. Therefore, important objectives of the ASP EE are the support of dynamic loading of new versions upon demand from protocol packets, while sharing a common byte code to reduce the memory footprint.

The ARP project has developed two significant active applications to execute under the ASP execution environment:

– *Jrsvp* implements the Internet signaling protocol RSVP. Except for framing, its network packet formats are compatible with the RSVP standard.

– *Jrip* implements the routing protocol RIP. This protocol is implemented as an AA but it sets the default forwarding table used by the ASP EE for ASP active packets. This is accomplished using a "Network Management Interface" (NMI) to the ASP EE.

3.4 ANTS

The Active Network Transport System (ANTS) [9] was one of the first active packet systems developed. Active packets, or capsules, logically contain the code that is needed to process them. Architecturally, ANTS is a Java-Virtual-Machine-based execution environment acting like a minimal NodeOS. The current ANTS prototype is written in Java and relies upon the JVM's bytecode verification and sandboxing facilities for the safety features they provide.

Local node resource usage is governed through the use of watchdog timers and memory allocation limits. The Java code is executed in the JVM and makes calls into a fixed and limited API to gain node-specific information. In addition, related capsules may leave a node-resident state for one another in a soft-state cache. Capsule types are grouped into protocols, and capsules are restricted to only access soft state belonging to their own protocol. The reference to the code actually takes the form of a MD5 cryptographic hash of the actual code preventing code spoofing. Thus, a misbehaving capsule is isolated from other capsules and the node itself, and if it consumes too many resources it is terminated. Furthermore, to control network-wide resource use, ANTS provides a TTL field, which is decremented at each hop, and duplicated when packets create a child packet. Since packets can create any number of child packets, the TTL limits the distance a packet's children can travel, but not the total work in the network.

One of the unique features of the ANTS system is an on-demand code loading system. Rather than carry the code itself, the capsules instead contain references to the code. If a node does not have a cached copy of the necessary code, it is loaded, typically from the previous node in the flow, but potentially from the source node. Thus the code for an active application marches 'ants-like' across the network as the caches are loaded. Perhaps the best demonstration of the flexibility of the ANTS system is the number and variety of active networking applications that have been built using the ANTS toolkit. These include specialized web caching, reliable multicast, mobile IP and distributed auction services [9].

Because the standard JVM does not support access to transmission resources at a sufficiently low level, implementations of ANTS on standard platforms cannot support quality-of-service capabilities, and are limited to the basic network capabilities provided by Java, and experiments have shown that the forwarding performance of capsules containing Java byte-code is dominated by overhead incurred due to execution within the Java virtual machine. However, efforts based on ANTS lead to the development of a new operating system called Janos [21]. Janos implements a special Java Virtual Machine and Java run-time for executing Java byte code. Janos includes a modified version of ANTS that supports resource management. Their main design goal is to provide a strong protection and separation mechanism between different user code executed at the active node. While execution of user code within a virtual machine (like Java) provides a high degree safety of execution, it also has associated performance costs. This is the time-honored approach of relying on the code supplier for safety in order to obtain better performance.

The aim of the Practical Active Network (PAN) [22] project is also to build a high performance *capsule*-based active node that is based on the ANTS framework. Architecturally, PAN is an in-kernel implementation of an EE using Linux as the node OS, able to saturate a 100 Mbps Ethernet with 1500 byte capsules containing native code and executed in-kernel.

3.5 Anetd

Anetd [23] is an experimental software designed to support the deployment, operation and control of active networks. It performs two major functions:

– Deployment, configuration, and control of networking software and distributed services, including active networking execution environment prototypes (e.g. ANTS), from a centralized source of electronic knowledge.
– Demultiplexing active network packets encapsulated using ANEP to multiple EEs located on the same network node and sharing the same input port.

The network services to be deployed are specified as URLs. Anted, after downloading the service with an HTTP GET command, strips the HTML header from the received code and installs the service. The current prototype supports the deployment of native binary compatible code or Java applications.

Anetd uses ANEP to support the merging of both the integrated and discrete approach to network programmability. Anetd not only accepts commands to implement the discrete approach (i.e. download, conFigure, etc.), but if the packet received is destined for a downloaded service, it forwards that packet to the appropriate service. Security is managed by using a 512-bit public key cryptography to authenticate control commands, and access control is managed through access lists and well-known servers.

3.6 The YAAP active platform

YAAP (Yet Another Active Platform) [24] was designed to be a simple prototype implementation of the mechanisms associated with active networks. YAAP active nodes are actually PCs running Linux OS as NodeOS using the ANEP protocol for demultiplexing active packets (Figure 7). A special *dispatcher* daemon is used to load and activate the services in the YAAP nodes. In order to provide a safe computation environment, the EE consists of a Java virtual machine in which the

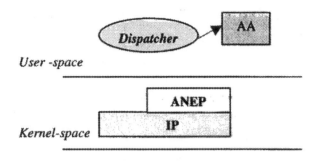

User -space

Kernel-space

Figure 7. *YAAP architecture*

services run, making use of Java Native Interface (JNI) to establish communication between the virtual machine in user-space and the kernel.

The dispatcher is a daemon running in user space and plays an important part of a YAAP node; it constitutes the interface between the kernel and the different components of the node, and serves also as a "bus" enabling the different components to communicate with each other as explained later. The dispatcher has also three main roles: the initiation of the download mechanism of a service, local creation of a service, and registration of a client to use a particular service.

Unlike other active platforms that use ANEP over UDP, ANEP is implemented within the kernel alongside IP, for obvious performance reasons. A new protocol code type specifying the ANEP protocol is used in the IP header. Intermediate YAAP nodes evaluate each ANEP packet even though the node is not the final destination. For that purpose, a minor modification was done to the IP routing code in Linux so that the kernel delivers the packet to YAAP, when the MSbit of the *Flags* field in the ANEP header is 0 allowing the packet to be "executed" locally.

YAAP supports loading-on-demand of services, i.e., when an ANEP packet arrives at a node with an unknown TypeID, and this packet is to be evaluated at that node, one role of the dispatcher is to initiate a download mechanism for the appropriate service, and launches the service. When a new service is launched at a node, it creates and binds to a YAAP socket, and listens on its own ANEP Type ID, thus permitting the kernel to deliver to it the appropriate packet. An API developed for YAAP (called *YAPI*) allows creation of ANEP sockets to send and receive packets in the same way as the TCP/UDP API.

This platform builds on concepts found in ANTS (e.g. demand-loading, Java execution environment) and Anetd (e.g. the use of ANEP). It also adds the possibility to download a service manually at any specified node in the network.

4. Conclusions

Viewing the network as a general computation engine enables an exciting set of capabilities and applications. Prior to the standardization efforts of the DARPA framework, early Active Networks platforms developed specific Execution Environments that made use of traditional machine operating systems for resources access and management. Porting execution environments to use standard NodeOS are currently being written, and is critical for refining the standards being developed to be able to reach a unified active networks architecture.

Current works are focusing on developing a common node OS. Here also, experimentation show a number of major open issues in active networking, including algorithms for integrated processor and link scheduling; algorithms for allocation of resources to flows within active nodes; and policies for instantiation and isolation of different abstract topologies.

One specific question currently debated is where to draw the EE/NodeOS boundary, and whether the NodeOS itself should be extensible, or if the ability to add functionality to the node should be reserved for the EEs. The answer seems to be that each NodeOS should be extensible in some way (i.e., analogous to loadable modules in traditional operating systems) and must allow the EEs to exploit these extensions, while not providing means for an EE to directly extend the NodeOS in order to be generic.

REFERENCES

[1] XIPENG X. *et al.*, "Internet QoS: A Big Picture", *IEEE Network,* March/April 1999.

[2] AMBROSH W., "The Intelligent Network", *A Joint Study by Bell Atlantic, IBM and Siemens,* Germany, 1989.

[3] "Open Signaling Working Group", http://comet.columbia.edu/opensig/

[4] BISWAS J. *et al.*, "The IEEE P1520 Standards Initiative for Programmable Network Interfaces", *IEEE Communications Magazine, Special Issue on Programmable Networks,* October 1998.

[5] CHAN M.-C. *et al.*, "On Realizing a Broadband Kernel for Multimedia Networks", *3rd COST 237 Workshop on Multimedia Telecommunications and Applications,* Barcelona, Spain, November 1996.

[6] CAMPBELL A. T. *et al.*, "A Survey of Programmable Networks", *ACM SIGCOMM Computer Communication Review,* Vol. 29, No. 2 p. 7–24, April 1999.

[7] CALVERT K. (Editor), "Architectural Framework for Active Networks", *DARPA AN Working Group Draft,* 1998.

[8] PETERSON L. (Editor), "NodeOS Interface Specification", *DARPA AN NodeOS Working Group Draft,* 1999.

[9] WETHERALL D. *et al.*, "ANTS: A Toolkit for Building and Dynamically Deploying Network Protocols", *IEEE OPENARCH'98*, San Francisco, CA, April 1998.

[10] WETHERALL D. *et al.*, "The ActiveIP Option", *7th ACM SIGOPS European Workshop*, 1996.

[11] ALEXANDER D. *et al.*, "Active Network Encapsulation Protocol", Draft, July 1997. Available at http://www.cis.upenn.edu/~switchware/ANEP/

[12] DECASPER D. *et al.*, "Simple Active Packet Format (SAPF)", *Experimental RFC*, August 1998.

[13] ZEGURA E., "CANEs: Composable Active Network Elements", *Georgia Institute of Technology*, http://www.cc.gatech.edu/projects/canes/

[14] CALVERT K. *et al.*, "Bowman: A Node OS for Active Networks", *IEEE Infocom 2000*, Tel Aviv, Israel, March 2000.

[15] Active Error Recovery (AER), http://www.tascnets.com/panama/AER

[16] DECASPER D. *et al.*, "A Scalable, High Performance Active Network Node", *IEEE Network*, January/February 1999.

[17] DECASPER D. *et al.*, "DAN – Distributed Code Caching for Active Networks", *IEEE INFOCOM'98*, April 1998, San Francisco.

[18] DECASPER D. *et al.*, "Router Plugins: A Software Architecture for Next Generation Routers", *SIGCOMM '98*, Vancouver, CA, September 1998.

[19] KELLER R. *et al.*, "An Active Router Architecture for Multicast Video Distribution", *IEEE INFOCOM 2000*, Tel Aviv, Israel, March 2000.

[20] Active Reservation Protocol (ARP), http://www.isi.edu/active-signal/ARP/

[21] "Janos: A Java-based Active Network Operating System", http://www.cs.utah.edu/projects/flux/janos/summary.html

[22] NYGREN E. *et al.*, "PAN: A High-Performance Active Network Node Supporting Multiple Mobile Code Systems", *IEEE OpenArch '99*, March 1999.

[23] RICCIULLI L., "Anetd: Active NETworks Daemon (v1.0)", http://www.csl.sri.com/ancors

[24] ROUHANA N. *et al.*, "YAAP: Yet Another Active Platform", *2nd International Workshop on Mobile Agents for Telecommunications (MATA)*, September 2000, Paris, France.

Chapter 6

Resource trading agents for adaptive active network applications

Lidia Yamamoto and Guy Leduc
Research Unit in Networking, University of Liège, Belgium

1. Introduction

In the context of an increasingly decentralized and heterogeneous network such as the Internet today, it is very difficult for the applications to know how much quality they can really expect. A lot of research effort has been dedicated to techniques to offer QoS guarantees, but these techniques are only effective when deployed in all the nodes concerned by a communication, or at least in all the nodes where resource shortage may occur. However, it is very difficult to achieve global agreements such that these techniques can be deployed. They are feasible in a private enterprise network, ISP or isolated operator, but not at the global scale. Therefore the only realistic answer seems to be to rely on applications that are able to adapt to a wide range of network conditions in a dynamic way, and in particular to the amount of resources available when these resources cannot be reserved in advance.

An adaptive application must be able to make optimal use of the available resources, and be able to adapt itself to fluctuations in resource availability. But this kind of application faces the difficulty of obtaining enough information about the network conditions, due to the Internet black-box model.

Clearly the use of more intelligent network elements in the network can allow applications to obtain the feedback they need to perform the adaptation functions more easily. Recently router support has been considered to assist adaptive applications to achieve better performance. But then the same question arises, how to add new functionality inside the network nodes without going through a long standardization process and without having to face the difficulties of global scale deployment.

Mobile agents and active networks can become useful tools to help in the adaptation process, since it becomes possible to inject customized computations at optimal points in the network, and the deployment problem can be easily solved using dynamic code mobility. In the long run this can lead to self-configurable, auto-adaptive network elements that are intelligent enough to "learn" the protocols they need to use at a given moment according to the devices available, services offered, operator policies, user demand, etc.

A special class of adaptation mechanisms is the so-called market-based control [CLE 96], for which a considerable amount of research results are available mainly in the agents field. It provides algorithms inspired by optimization and economy theories for distributed control of resource usage, with many applications to computer and telecommunication networks. The benefit of such mechanisms is two-fold: on one side, optimal resource sharing configurations can be achieved in a decentralized way; on the otherside, it becomes easier to quantify heterogeneity in terms of resource availability, to offer the users the opportunity to trade one type of resource for another.

However, relatively few results have been shown which directly apply such artificial economy models to the specifics of active networks, with special attention to highly adaptive applications. We address this issue in this article by providing a simple model which allows the active applications to make decisions about the amount of resources to use, according to the network conditions found in the active nodes. Using such a model, an audio mixer is developed as an instance of adaptive active network application, which is able to trade bandwidth for memory according to the available prices of each resource.

The article is organized as follows: in Section 2 we review the state of the art in agent and active network techniques for adaptive applications. Section 3 presents our model to trade resources inside an active node. Section 4 applies the model to the case of an audio mixing application. Section 5 shows simulation results for the audio application, and Section 6 concludes the article.

2. Background

In this section we give a survey of the current research directions concerning agent and active network techniques applied to adaptive applications. We start with an overview of current techniques used in network and transport level adaptation protocols, and then discuss the potential of mobile agents and active networks for such applications, with a survey of current proposals in this area. After that we review market-based control research applied to resource sharing in computer and telecommunication networks. The use of such techniques in the active network context is the focus of our work.

2.1 Adaptive applications

Adaptive applications can tolerate fluctuations in resource availability, and are necessary in a heterogeneous environment such as the Internet today, where different network technologies and user terminals are interconnected together, and over which a multitude of services coexist. In the case of multimedia applications, a good survey can be found in [VAN 00].

The adaptation mechanisms can be implemented at several layers of the protocol stack, ranging from pure application layer techniques to network level protocols. For example, we can adapt to the available bandwidth using elastic traffic that reduces the data rate generated in presence of network congestion. Fluctuations in delay can be dealt with by using elastic buffers to adjust the play-out time. To deal with CPU and memory bottlenecks, some interaction with the operating system is necessary (see Section 5 of [VAN 00] for examples). Our focus in this paper is on network and transport level mechanisms for adaptation. At this level, the classical approaches are typically end-to-end based and adapt to the available network bandwidth (congestion control). The best known example is the TCP protocol, which adjusts its window size in response to the current level of congestion. Since TCP is so widespread today, but does not support the emerging multimedia applications, a lot of research is in progress in order to obtain TCP-friendly protocols for real-time and streaming applications, for unicast as well as multicast traffic [FLO 00] [KHA 00].

It is well known that classical end-to-end approaches often present drawbacks such as annoying quality fluctuations, sub-optimal resource utilization and sharing, and slow convergence. Part of these problems come from the fact that the applications try to adapt in a blind way, without having enough information about the network conditions, since the Internet uses a black-box model, which hides all information from the end hosts.

Recently some router support has been considered to assist adaptive applications to achieve better performance, e.g. [GOP 00]. However, how to have such schemes widely accepted and deployed is still a question mark.

We believe that code mobility as provided by mobile agents and active networks is the only generic solution to allow adaptive software to be incrementally deployed and evolve through usage experience. In Section 2.2 we give a brief introduction to the active network concepts used throughout the article, and in Section 2.3 we discuss the relationship between mobile agents and active networks. In Section 2.4 we review the current research efforts towards adaptation protocols that can benefit from code mobility.

2.2 Active networks

Active networks (AN) allow the network managers or users to program the network nodes according to their needs, offering a great amount of flexibility. The nodes of an active network [TEN 97] are capable not only of forwarding packets as usual but also of loading and executing mobile code. The code can be transported out-of-band, within specialized signalling channels (programmable networks) or in-band, within special packets called "capsules" (active networks). Capsules might contain the code itself (such as in [HIC 99]) or a reference to it, such that it can be downloaded when the first capsule containing the reference arrives at a given node (such as in [WET 98]).

If the distinction between active and programmable networks seemed at some point in time clear [TEN 97] [CH 98], the tendency today seems to be towards an integration of the two concepts [ALE 99], since both are forms of achieving open programmability in networks [CAM 99], and special flavours in between or combining both approaches are also possible [ALE 99] [HJA 00]. In this paper we focus on the capsule model in order to be more generic and avoid programmability restrictions.

There are basically two AN architectural lines, derived from the two original communities on out-of-band and in-band programmability. The first one is the IEEE P1520 reference model [DEN 99], which offers a set of standard interfaces to program IP routers, ATM switches and other network devices. The P1520 model is the first standardization effort towards open network programming interfaces, and is likely to be one of the first AN interfaces to become commercially available.

The second architectural line is the framework for an active node architecture [CAL 99] which is being proposed within the DARPA AN research community. It includes a supporting operating system (the NodeOS), one or more execution environments (EE), and the active applications (AA). These components are outlined in Figure 1. The NodeOS is responsible for managing local resources

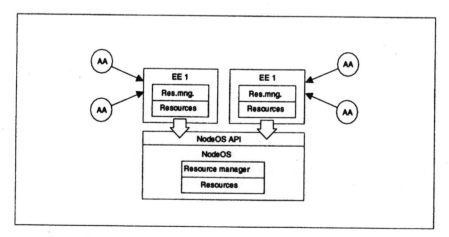

Figure 1. Main components of the DARPA AN architecture

such as CPU processing time, link bandwidth and memory storage. On top of the NodeOS, a number of EEs can be installed. On top of each EE, various AAs can be dynamically loaded and executed. The EE is responsible for controlling the access from the AAs to local resources, and limiting resource usage depending on specified policies.

The NodeOS plays a crucial role in providing access to local node resources, as well as information about resource availability. A NodeOS API is currently being defined [PET 00]. At the moment this API treats four types of resources: computation, memory, communication, and persistent storage. The communication resource is handled through the channel abstraction, which when ready should include QoS support, as well as access to link information such as bandwidth, queue length, and other properties and statistics.

2.3 Mobile agents and active networks

Mobile agents are autonomous pieces of mobile code that travel through the network acting on behalf of their owners. There is a vast amount of literature on the subject, applied to several areas such as manufacturing, e-commerce, network management [HAY 99a] [PAP 00].

The intersection between mobile agent technology and active network technology is the use of mobile code. The capsules of an active network can be seen as subclasses of mobile agents, specialized for network-related operations. The similarity between active networking and mobile agent concepts is briefly discussed in [HAY 99b] [BRE 99].

Actually, we can consider that the pioneer AN platform came from a mobile agent framework: the MØ platform [TSC 93]. MØ is based on the concept of messengers as mobile computational entities that are able to perform any network service. In [TSC 93] it is shown that any protocol based on the classical PDU paradigm – which is the case of basically all the network protocols in operation so far – can be implemented using the messengers paradigm. On the other hand, it is also pointed out that not all the protocols that can be implemented with messengers can be implemented using simple PDUs. The typical case is the one of protocols that are able to evolve their own code on the fly. The author called such protocols "genetic protocols" but admitted to be unable to come up with concrete examples where such protocols would really be necessary in practice. Since then, the search for such protocols still persists in the AN and mobile agent worlds [BOL 00].

If conceptually the border line between mobile agents and AN capsules or messengers is a blur, in practice many differences subsist between the two approaches. Mobile agents concentrate mainly on application-level or network management duties, and can typically accomplish much more complex tasks, with richer functionality, that what is generally allowed to capsules or even to out-of-band active code. Moreover, while capsules can be designed to be autonomous and mobile, downloadable modules for programmable networking are typically like plug-ins, and have no autonomy nor mobility once they are installed at the target network element. The architectures for mobile agent platforms and active network platforms differ in the kind of support for code mobility that is offered. Mobile agent platforms tend to concentrate on application level or network value added services, while active network platforms are optimized for transport rather than processing of information.

It is interesting to note that the simplicity of AN platforms compared to agent platforms may enlarge their applicability beyond the network domain they were initially designed for. An example of that can be found in [RUM 00], where the concept of "active network calls" is introduced, which uses AN capsules to support distributed computations. The authors report that efficiency was the reason for the choice of AN capsules instead of mobile agents.

On the other hand, mobile agents can also be used as enabling platform for active networks as pointed out in [BRE 99] [SUG 99]. In this case, instead of the usual plug-ins, it is possible to benefit from full mobile agent functionality in an AN environment.

Finally, it is important to remember that both mobile agent and AN technologies face the same challenges which are mainly security, performance, resource management and interoperability.

2.4 Resource management

One of the main difficulties encountered in classical adaptation approaches is how to obtain the required information about resource availability, mainly when this information is hidden in a black box network and has to be inferred using only some indirect indications that are observed at the end systems. Using code mobility as provided by mobile agents and active networks, new models for adaptive applications could be envisaged, which can benefit from the possibility to send capsules or agents to certain elements inside the network. These agents can be in charge of collecting information about network conditions, without having to rely on indirect indications or on heavy signalling protocols. Indeed, the idea of sending small pieces of code directly to where the data needs to be treated, instead of exchanging a large amount of data, is one of the main motivations of mobile agent technology, and it can also be applied to mobile code in the case of active networks.

Actually many adaptation mechanisms come from the world of mobile agents. In [JUN 00] an adaptive QoS scheme for MPEG client–server video applications is described. It is based on intelligent agents that reserve network bandwidth and local CPU cycles, and adjust the video stream appropriately. Many agent-based adaptation schemes use artificial market mechanisms [CLE 96]. In [YAM 96] a market model to allocate QoS is applied to a conferencing tool targeted at casual meetings where sudden variations in bandwidth availability require an adaptive QoS control strategy. In [TSC 97] an open resource allocation scheme based on market models is applied to the case of memory allocation for mobile code. [GIB 99] describes a market-based mechanism to set up circuit switching paths with resource reservation. In [BRE 00] the problem of budget planning for mobile agents is addressed, such that they can successfully complete their tasks given their limited budget constraints.

In the domain of end-to-end congestion control in networks, optimization schemes (see [LOW 99] for an overview) are receiving a lot of attention recently. The algorithms derived from such studies also use a price measure to indicate the level of congestion, and utility functions to quantify users' share of bandwidth. They are therefore closely related to artificial economy systems, but they are restricted to the specific case where bandwidth is the only scarce resource.

Active networks can more easily benefit from the various adaptation mechanisms described, when compared to classical networks, since their code can be dynamically deployed. However, most of the current AN architectures still offer little support for such mechanisms to be implemented in a straightforward manner. For example, ANTS capsules carry a resource limit field that is decremented for the consumption of resources [WET 98]. When it reaches zero the capsule is discarded. However, no mechanisms are specified to manage this field, or to quantify the amount to be decremented.

This situation is rapidly changing though. Recently, [ANA 00] presented a very promising market-based resource management infrastructure for active networks. It includes a distributed trust-management system which ensures that the market-based policies can be properly enforced in a scalable way.

Another step towards market-based models for AN is the cost model proposed in [NAJ 00], which expresses the trade-off between different types of resources in a quantitative way. However, the recursive approach adopted makes its use more appropriate in the context of reservation-based applications, instead of highly adaptive ones.

A crucial issue for resource management in active networks is the support for incremental deployment. The success of active networks, protocols and services will depend on their ability to complement and interoperate with existing networks in a transparent way, such that active nodes and active functionality can be incrementally deployed. In a hybrid network where only a few nodes might be active, it is important to be able to estimate the resource availability outside an active node. In [SIV 00] an *equivalent link abstraction* is proposed as part of the Protean architecture. Using this abstraction, it is possible to consider a set of non-active nodes as a single link from the point of view of the active nodes involved. Such a virtual link presents changing properties which must be discovered in real-time, such as average rate, delay and packet loss probability. The use of this type of abstraction in the context of artificial market models, as well as its extension to multi-access links, to the case of asymmetric and/or changing routes, etc., are still open issues.

3. Resource trading model

In previous work [YAM 00a] we proposed a model for trading resources inside an active node, and applied it to an audio application. This paper is an extended, corrected and updated version of that work. The model was derived from our earlier work on a layered multicast protocol [YAM 00b], therefore it is not a mere abstract proposal but is motivated by concrete application needs.

Our model aims at offering a generic communication abstraction between active network agents, such that different adaptive applications and different resource management policies can be implemented. We are mainly concerned with adaptation to available resources when resources cannot be reserved in advance, either because the router itself does not support reservations, or because the network is heterogeneous and some of the routers along the path offer no QoS support.

In the model, two types of agents communicate to seek an equilibrium. Each agent tries to optimize its own benefits: on one side resource manager agents have the goal of maximizing resource usage while maintaining a good performance level. On the other side, user agents try to obtain a better quality/price relation for the resources consumed, and to efficiently manage their own budgets avoiding waste. Both types of agents are implemented as AAs with different privileges, and they communicate such that the resource managers can "sell" resources to the user agents at a price that varies as a function of the demand for the resource.

A currency is introduced into the system to allow for trading of different resource types. This is the basic requirement for artificial economy models such as [BRE 00] [FER 96] [TSC 97] to develop in active nodes, and is also an essential feedback parameter for most algorithms based on optimization (e.g. [LOW 99]).

The idea is to enable auto-configurable applications and resource managers, such that the code from both types of agents can be dynamically loaded, in order to make them evolve to adapt to new conditions. Figure 2 shows the model as it could be implemented over the DARPA AN architecture. The main implementation difficulty at the moment seems to be the definition of interfaces to local information concerning resource availability, which need to be exported to the active applications. In this matter, we can learn from the mobile agents field in order to model the interactions between resource manager agents and user agents. Software agent communication paradigms such as Agent Communication Languages can be helpful, but still need to be specialized to the AN context.

Such a model is per se not entirely new, and is in fact a mere simplification of existing artificial economy models which have been mainly applied to the agents

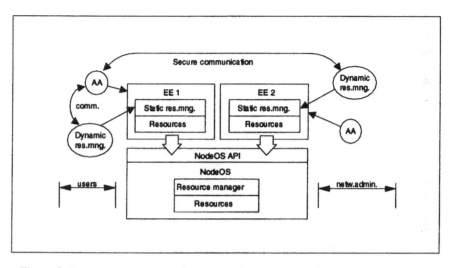

Figure 2. *Resource managers and active applications. Hypothetic placement over the DARPA AN architecture*

domain. For example, the scheme in [GIB 99] is similar to ours, although more complex. The main difference with respect to our model is that we try to adapt it to the specifics of adaptive applications over active networks (no resource reservations), in which reaction time is critical, therefore precluding the use of complex transactions.

The AN market-based infrastructure proposed in [ANA 00] allows the trading of resources between producing agents and consuming agents, which can be mediated by service broker agents. The currency used takes the form of a resource access right. This architecture also bears many similarities with our model. The main different is that we focus on AAs that can trade resources, while [ANA 00] focuses on the underlying architecture and security mechanisms to support such AAs. Our research efforts are therefore complementary, and we plan to implement our ideas over the platform developed by [ANA 00].

3.1 Resource manager agents

Resource managers export resource prices which are a function of the resource utilization. The utilization is related to the load, and to the demand for a resource. The function or algorithm used to calculate prices can be shaped to implement desired policies, such as to achieve high utilization, but also to offer good quality to the users. Resource managers may contain dynamic and/or static code. Some lower level functions which are especially time-critical might be implemented using static code (or even in hardware) while the mobile part would be used to implement more complex policies and to select from a set of pre-existent lower level functions. It is necessary to have the possibility to use dynamic code in order to be able to improve strategies, that is, make them evolve over time, in an active network. Here is one of the places where the alliance between AN and mobile agents can become a must: resource managers can be deployed using mobile agents that are sent by the network manager in order to install new policies. Classical mobile agents for network management can be used for this purpose.

Resource managers are implemented as AAs injected by the network provider, which are executed with network administrator's privileges. There is one class of resource manager for each type of resource concerned. The most relevant classes are: link managers, CPU managers, and memory managers. We will give some more attention to link managers, since they play a crucial role in network congestion control.

Link manager agents also allow abstractions to be made which enable the user applications to adapt to a wide variety of environments in a transparent way. By exporting prices instead of the link internal state information directly, it is possible to hide the specific details of link characteristics while at the same time offering a proper congestion indication to adaptive applications. For example, the price function for a classical point-to-point link would be different from that of a multiple access link, where the local interface load is not a good indication of the actual link utilization.

An important feature of the link manager price abstraction is that it allows the active applications to deal with non-active nodes in a transparent way. For example, a link manager that implements the equivalent link abstraction [SIV 00] could export a price which would be a function of the estimate average rate, average delay and packet loss probability, which are changing properties in the case of an Internet virtual link. As discussed earlier, this type of abstraction is crucial for the success of active networks, since the deployment of active nodes will depend on their ability to complement and interwork with existing technologies.

A lot of research is still to be done on how to adjust prices in artificial agent-based economies. An interesting analysis can be found in [MIZ 99]. For this paper, however, this will not be our focus. We will rather concentrate on the user side, assuming that the resource manager agents in the active nodes are able to implement suitable pricing algorithms.

3.2 Capsules as reactive user agents

The second type of agent in the model are the active packets themselves, modelled as simple reactive mobile agents. Actually capsules can be regarded as a specialized subclass of mobile agents. They travel to network nodes where they decide when to continue or stop the trip (e.g. stop due to congestion), when to fork new capsules (e.g. in a multicast branch), and which amount of resources to use at each node in order to complete their (generally simplified) tasks.

Capsules have the properties of autonomy and mobility, but need to be simple enough to be executed at the network layer, where performance is often critical. They need to take fast decisions using little data and reasoning, therefore they must be extremely reactive. They can therefore be classified in the reactive agents category. This does not exclude from the model the possibility of using more intelligent mobile agents, jointly with capsules. However, in this paper we focus only on capsules.

Capsules generally represent the user interests, with the goal of attaining the best possible quality at the lowest possible cost. This means that a capsule must be able to make rational decisions on how to spend its limited budget, after consulting resource manager agents for information about the prices of the various resources needed.

Capsules carry a budget that allows them to afford resources in the active nodes, as the resource limit field in ANTS [WET 98]. It looks like a TTL (Time-To-Live) field which is decremented by a number of units for the consumption of a certain amount of resources. When it reaches zero the capsule is discarded. As explained in [WET 98], the budget field must be protected such that the capsules themselves cannot modify it. Again, since capsules are reactive agents, the transactions must be kept simple enough. Instead of the auction mechanisms frequently used by agents [CLE 96], the capsules will most of the time either accept or refuse to pay a given price. Refusal might imply that the capsule simply disappears from the system, because it does not have enough budget to proceed on its journey to its destination. Again, we do not preclude the usage of more complex mechanisms implemented by intelligent mobile agents. These agents can also have access to the resource manager interfaces; therefore they can also benefit from the model, but they are not our focus as pointed out earlier in this section.

At the end systems (hosts) conventional programs or intelligent agents can be used to spawn capsules. Since these hosts can dedicate significant amounts of resources to information processing, sophisticated strategies can be used to decide which capsules to spawn and when, and how much of the total user's income can be assigned to each capsule (planning). These tasks are then not delegated to the capsules themselves, but can be delegated to more generic mobile agents such as the ones described in [BRE 00].

Here we face the problem of how the budget should be distributed between capsules for one user, and among different users, and also of how to make purchase decisions according to budget constraints. This can be done with the help of utility functions, which quantify the level of user satisfaction for receiving a certain amount of a good, or more generally a combination of goods (market basket) [PIN 98].

According to economics, a typical behaviour of a rational consumer agent is to try to maximize its utility subject to the budget constraints given by its limited income.

For an example where only two goods are involved, the user optimization problem is then typically expressed as:

Maximize U(x,y)

subject to: $p_x \cdot x + p_y \cdot y = 1$

where $U(x,y)$ is the utility function, x and y are the quantities of two goods in their respective units, p_x is the price per unit of x, p_y is the price per unit of y, and I is the user's income.

Such a maximization process will lead to an equilibrium if the utility function satisfies some properties such as being strictly concave increasing, i.e. the increase in satisfaction is smaller the more items of one good are consumed. The increase in satisfaction that a user obtains from consuming one more item of a given good is called the marginal utility. Within the rational consumer assumption, the marginal utility is a decreasing function of the number of items of a given good. This is called the diminishing marginal utility assumption, and it is directly related to the demand function of a given user for a given good. It means that the more one has from something, the less it is willing to pay to obtain more of it.

A typical utility function is the Cobb–Douglas utility function, given by:

$$U(x,y) = a \cdot \log(x) + (1-a) \cdot \log(y)$$

where $0 \le a \le 1$ is a constant which represents the importance the user assigns to x with respect to y.

Applying the method of Lagrange multipliers, the solution of the user optimization problem with the Cobb–Douglas utility function is given by [PIN 98]:

$$x(p_x) = \frac{a \cdot I}{p_x} \qquad\qquad [1]$$

$$y(p_y) = \frac{(1-a) \cdot I}{p_y} \qquad\qquad [2]$$

The functions $x(p_x)$ and $y(p_y)$ are the demand functions for goods x and y given their current prices per unit p_x and p_y respectively. With demand functions shaped like these, it is possible to choose the quantity of a given resource to consume in order to maximize user satisfaction, given the current price for the resource and the agent's income. Since the price information for each type of resource is available in the active nodes, it is possible for an agent to calculate the amount of resources it can consume at each node, given only an upper bound on the budget per hop it has planned to spend. In the case of capsules, the planning decisions to calculate this upper bound could be performed either at the hosts or by more intelligent mobile agents that would be sent to the active nodes less frequently than capsules. The techniques for such planning are out of scope of this article, and the interested reader can refer to [BRE 00] for more information.

The same reasoning can be easily generalized to an arbitrary number of resources [FER 96]. However, for the purpose of our study this will be sufficient, since we will restrict ourselves to link and memory resources.

Another important characteristic of the Cobb–Douglas function is that it allows us to easily quantify the preferences of different consumers towards one good. For example, given two user agents u_1 and u_2 with the same income I, if agent u_1 has the weight a_1 on its utility function for good x, and agent u_2 has weight $a_2 = n{\cdot}a_1$ for the same good, we have:

$$x_1(p_x) = \frac{a_1 {\cdot} I}{p_x}$$

$$x_2(p_x) = \frac{a_2 {\cdot} I}{p_x} = \frac{n {\cdot} a_1 {\cdot} I}{p_x} = n {\cdot} x_1(p_x)$$

Thus for a given price p_x we have:

$$a_2 = n {\cdot} a_1 \Rightarrow x_2 = n {\cdot} x_1$$

It means that if we know that agent u_2 values resource x twice as much as agent u_1, then when faced with the same price, u_2 will get twice as much of x as u_1. This is an interesting tool for computer network applications, since it provides a quantitative way to provide service differentiation according to the preferences of users. This capability is already known from literature, e.g. [LOW 99] [FER 96]. For example, when trading bandwidth for memory storage, a time-constrained application such as an audio-conference will certainly prefer bandwidth to storage if prices are the same, while a bulk transfer application would probably prefer to store as much information as possible when the links are congested, in order to avoid losses and retransmissions.

In real world economies it is difficult to quantify utilities, but in artificial economies this might be less difficult. Without having to relate artificial currencies to real ones, we could imagine that the maximum budget per unit of time is controlled by a policy server from the network provider that the user is subscribed to, in order to guarantee that users will employ such budget rationally and prevent malicious users from grabbing most of the resources by marking all their capsules with a high budget. This can be compared to the IETF diffserv policies. However, in diffserv only a few predefined classes of service are available, while with such artificial economies, a whole range of classes could appear (and eventually die out), defined by their particular utility characteristics.

4. Congestion control for a concast audio mixer

We illustrate the use of the trading model through a congestion control scheme for a many-to-one (concast) service. The concast example shows capsules that trade brandwidth for memory when there is congestion.

The term "concast" has been defined in [CAL 00] as a many-to-one service, in opposition to multicast (one-to-many). Figure 3 illustrates this concept. While multicast copies information from one source to many destinations, concast merges information from several sources to one destination. A concast service can be used, for instance, to aggregate feedback in a reliable multicast service, to transmit reception statistics in a multimedia session, to merge information coming from several sources in an auction or tele-voting application, or to combine several real-time streams into one, e.g. to perform audio mixing from several audio sources.

The concast service is faced with the feedback implosion problem, that is, multiple simultaneous sources might congest the path to the single destination, if no congestion control is performed. This problem is aggravated by the fact that many concast flows are used as signalling to support a more robust protocol such as the aggregation of NACK feedback messages for a reliable multicast protocol. Flow control for such signalling messages is often neglected or oversimplified, since the signalling traffic is assumed to be kept small enough when compared to the data traffic. This might lead to poor performance when there is congestion in such a signalling path. We argue that for future sophisticated active services congestion control will be equally important on the signalling and data paths, since the distinction between both tends to become naturally blurred as we approach new network service composition frameworks which are not simply stack-based as the classical OSI model.

For the purpose of this study, we focus on the case of an audio mixing application. However, the same ideas are applicable to any application making use

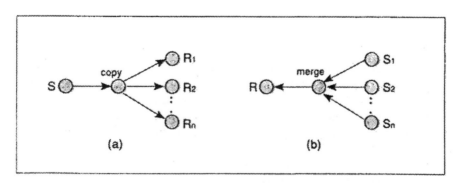

Figure 3. *Multicast (a) versus concast (b) service abstractions*

of the concast service, differing only in the way that packets are combined (merge semantics).

In our audio mixing application, several sources generate audio streams in a session. The streams are collected at a single node, which can either record them, play them back or redistribute them to a multicast group. If the receiving node collects all the data and mixes it locally, it might end up with the implosion problem. It is possible to perform mixing operations at every active node, so that the receiver gets a single stream already mixed. However this might cost too much processing and/or memory space in the active routers, so it might become too expensive. Also note that mixing streams delays them, since it is necessary to wait until a minimum number of packets arrive in order to sum up their audio payloads.

The mixing operation is a physical sum of audio samples. Therefore the resulting audio payload has the same number of bytes as each of the original payloads. We assume that there is a maximum amount of signals that can be added up without saturation. Another assumption is that a constant bit rate code is used, with no silence suppression, such that the signals are generated at a constant rate from the beginning to the end of the session. This initial rate will be altered along the path, as capsules are mixed according to the network conditions.

Note that the mixing operation requires a list of addresses, in order to identify the list of sources already mixed. One might argue that such a list might also occupy bandwidth, since the packet size has to increase in order to accommodate it. But if an estimation of the group size is available, a fixed-size bitstream can be used to hold the list of sources. Adding or removing a source becomes as simple as setting or resetting a bit in the bitstream. The intersection and union operations can also be implemented as AND and OR binary operations respectively.

We propose to trade bandwidth for processing and memory space to achieve a compromise solution which uses resources efficiently and therefore is able to control congestion. The idea is that when there is congestion at an outgoing interface, the consequently high bandwidth prices will push the users to save bandwidth by performing mixing of data. When the congestion clears up, the users can again benefit from the available bandwidth to avoid the extra delays imposed by the mixing operation.

When a capsule arrives at a node, the first thing it does is to check whether another payload coming from the same source is already buffered. In this case, the arriving capsule cannot mix its payload to the buffered one, which carries earlier samples. Note that this check is in fact an intersection operation to check whether one of the sources mixed in the current capsule is already buffered. If that is the case, the arriving capsule immediately dispatches the buffered payload by creating a new data capsule and injecting it into the local execution environment. This procedure also prevents misordering of packets. As a consequence, at most one packet payload per session is buffered in an active node. We assume that all payloads have the same size.

After that, the capsule can take a decision to either proceed, buffer its payload (for mixing), or discard itself. This decision depends on the budget it carries, and on the prices of the memory and link resources it needs. A number of alternative decision strategies can be envisaged:

Null strategy: Corresponds to the trivial case when no congestion control is performed. In this case the capsule always goes intact to the outgoing interface.

First strategy: If there are other audio payloads from the same group waiting in the memory buffer, it adds its own payload to the buffered one and adds its list of sources to the list of sources already buffered (union operation). Then it terminates execution. If no other audio payloads from the same group are buffered yet, it decides for the cheapest resource: if the price of memory (to buffer the payload for future mixing) is currently lower than the transmission price for the capsule, then it decides to buffer itself; otherwise it decides to move on to the next hop.

Second strategy: The arriving capsule mixes its own data with the buffered one (if existent), then it chooses the cheapest resource, either memory or bandwidth.

The difference between the first strategy and the second strategy is that the first strategy always decides for storage when another payload from the same session is already buffered, while the second strategy always chooses the cheapest resource, independent on the fact that another payload is already buffered or not.

Although the first and second strategies are very naive, they already give quite reasonable results as we will see below. However, their behaviour is sub-optimal and they do not take into account the different preferences for resources.

Third strategy: It first mixes its own data with the buffered one (if existent), as in the second strategy. It obtains a new payload that combines samples from n sources (n is known from the list of source addresses). It also knows N, the maximum number of sources that can be mixed together without saturation. Then it calculates the amounts of link and memory resources according to equations [1] (for link) and [2] (for memory), where the a parameter is also carried in the capsule, and expresses its preferences for link resources with respect to memory. It then tries to keep the resources in the proportions obtained, as follows:

If $\dfrac{N-n}{n} > \dfrac{x}{y}$ then store, else move on,

where x is the demand for link resources according to equation [1], and y is the demand for memory resources according to equation [2].

Since we have:

$$\frac{x}{y} = \frac{a \cdot p_y}{(1-a) \cdot p_x}$$

the decision is independent on the capsule's budget I. The a parameter will play a role in the proportion of link resources used with respect to memory resources. The higher a, the higher the amount of link resources used, and therefore less capsules will be mixed together, for given memory and link prices.

Note that in all cases, capsules that run out of budget are automatically discarded by the resource managers, thus there is no need to explicitly indicate this operation.

5. Simulations

The audio mixing AA has been simulated with the help of an AN module that we developed for the NS simulator [NS 00]. This module implements a simplified AN architecture consisting of a NodeOS, an EE, and some resource managers. The simulated EE executives capsules written in TCL language.

The topology for the simulations is shown in Figure 4. It consists of n sessions of m sources and one receiver each. The sessions traverse a bottleneck (link L), so that the capsules in active node N must decide to mix or to proceed intact to the receiver node, according to the prices of link or memory resources available from the resource managers.

The price function used is based on the one in [TSC 97]:

$$price = 1000 \cdot \sqrt{\frac{1.01}{1.01 - load}} - 1000$$

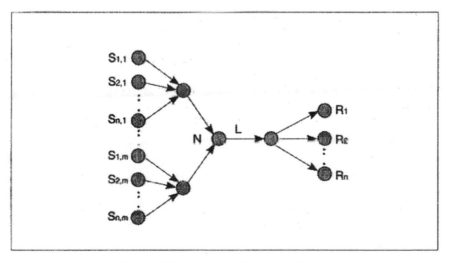

Figure 4. *Topology used in the simulations*

This function is a practical implementation of a convex increasing function that would go to infinity as the load approaches 100%. It forces the price to rise sharply as we approach high loads, which discourages applications from using a resource when its load is too high. This gives the applications a clear indication of the "dangerous zone" to avoid, while at the same time encouraging a relatively high utilization.

In the case of the memory manager, the load is given by the ratio between the average number of memory units occupied, and the total number of memory units available to user capsules (which is of course assumed to be much smaller than the actual amount of memory available). The average is calculated using an exponential weighted moving average (EWMA [FLO 93]).

For the link manager, the load is given by the average queue occupancy ratio at the outgoing link interface. This average ratio is obtained by calculating the average queue length as an EWMA, and then dividing by the maximum queue size. The resulting congestion indication is a bit similar to RED, except that here the binary feedback is replaced by an explicit price indication to the arriving capsules. Note that the actual usage of bandwidth is not taken into account in the price function, if it does not cause queues to build up. This is therefore a very simplified version of a link manager, but it already serves the purpose of controlling congestion.

We first run an example where 2 sessions are active. There are 5 sources per session, each sending an audio stream of 100 kbps to a single receiver, resulting in a total of 1 Mbps of traffic arriving at N. The capacity of link L is set to 500 kbps.

Figure 5 (top) shows what happens in the trivial case when no congestion control is used. In this case, link overflow occurs at L, and half of the packets are dropped. The remaining packets receive and unequal share of the bottleneck link as we can see in the figure.

Figure 5 (middle) shows what happens when the first strategy is used. First we can notice that the two sessions (left and right) get approximately the same share of the bottleneck. Additionally, no packet losses were observed during the simulation. However, the link is underutilized. This can be explained by the fact that this strategy always favours memory when there is already an item in memory.

The second strategy is a bit more clever (Figure 5, bottom). It always chooses the cheapest resource, either memory or bandwidth. Therefore it is able to grab any bandwidth when it becomes available. Here again, no packet losses occur. However, with this strategy it is not possible to specify different weights for each resource.

Now let us look at the third strategy, which allows us to specify utility weights through the a parameter. Figure 6 (top) shows its behaviour when the weights of the two sessions are the same. We see that this strategy is able to share the bandwidth efficiently. It also leads to more stable rates when compared to the previous strategies. The same happens when the weights are different (bottom side

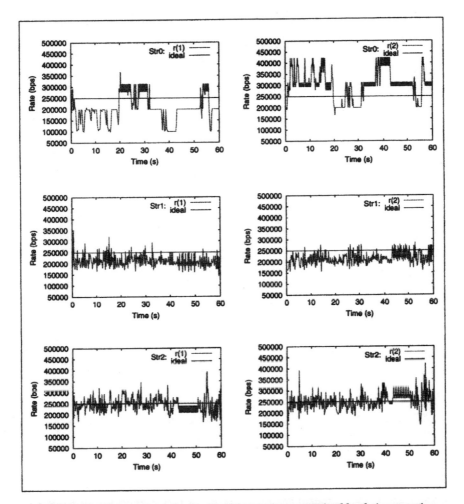

Figure 5. *Evolution of rates in time for 2 sessions, as perceived by their respective receivers. Left: first session (receiver r_1). Right: second session (receiver r_2). Top: no congestion control. Middle: first strategy. Bottom: second strategy*

of Figure 6), but in this case each session receives link resources in proportion to its respective weight, expressed by the a parameter value.

Until this point only the rates have been shown, since our main goal is to achieve congestion control. Table 1 shows average memory and link parameters taken over the complete duration of the simulations shown in Figures 5 and 6. The null strategy (Str.0, no congestion control) occupies most of the link resources and causes the link price to rise, since the link queue is most of the time full. Strategy 1 reduces link utilization by using memory for mixing, however it does that in an

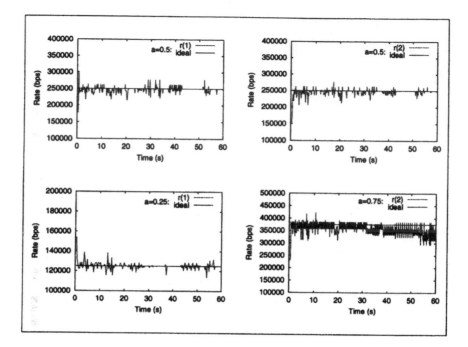

Figure 6. *Evolution of rates in time for 2 sessions using the third strategy.*
Top: $a_1 = a_2 = 0.5$ for link resources. Bottom: $a_1 = 0.25$, $a_2 = 0.75$

inefficient way when compared to strategy 2, which is able to use more bandwidth while decreasing both link buffer occupancy and memory utilization. The result is a decrease in link and memory prices.

Strategy 3, when $a_1 = a_2$ (Str.3a in Table 1), improves further by keeping the average link queue occupancy at a very low level, in spite of a high bandwidth utilization. This can be explained by the fact that this strategy tries to find an optimum balance between the usage of memory (which delays packets) and the usage of link resources. It therefore waits for the good moment to send a packet over the link, the movement when link prices are favourable (which in our case corresponds to low queue occupancy).

Column Str.3b in Table 1 shows the average resource parameters when $a_1 \neq a_2$, corresponding to the situation illustrated at the bottom side of Figure 6. There are no significant changes in the parameters at node N, with respect to Str.3a, but the prices are higher, which is an unexpected result, as we we would expect the two sessions to concentrate each one on its respective preferred resource, making the load equally distributed. We are still investigating the reason for this discrepancy.

Table 1. Memory and link usage parameters for different strategies

Parameter (average)	Str.0	Str.1	Str.2	Str.3a	Str.3b
memory utilisation (%)	0	7.18	5.14	4.98	5.33
bandwidth utilisation (%)	99.91	84.54	98.18	98.93	96.37
link queue occupancy (%)	93.58	5.83	4.97	1.61	2.98
memory price ($)	0	37.71	26.57	25.62	27.37
link price ($)	2914.6	32.35	27.71	10.17	17.22

In order to have a better insight on the mixing procedure, we now look at the number of audio sources carried by each capsule that arrives at its destination. Table 2 shows the percentage of packets that arrived at both receivers, over the total number of packets sent, which is the mix of a given number of sources. The first row ("0 (lost)") represents lost packets. The second row (one source) represents the percentage of packets that arrive intact from the source. The third row (two sources) represents packets that mix samples from two different sources, and so on for the rest of the rows. The columns represent the strategies used. For the null strategy, roughly half of the packets are lost, and the remaining packets arrive intact (no mixing). Strategy 1 is an all-or-nothing strategy: two thirds of the arriving packets contain data from only one source, while one third contains data from all the sources for a given session. The second strategy distributes the mixing effort more evenly.

Table 2. Average percentage of packets carrying the sum of samples from a given number of sources

# sources	Str.0	Str.1	Str.2	Str.3a	Str.3b(r1)	Str.3b(r2)
0 (lost)	49.61	0	0	0	0	0
1	50.39	66.22	39.87	1.54	0.05	60.86
2	0	0.02	33.53	96.16	0.48	39.14
3	0	0	15.80	2.30	1.93	0
4	0	0.43	6.26	0	97.54	0
5	0	33.34	4.53	0	0	0

As for the third strategy, column Str.3a of Table 2 shows the results when $a_1 = a_2$, corresponding to the simulation result shown at the top side of Figure 6. Columns Str.3b(r1) and Str.3b(r2) show the results for receivers r_1 and r_2 respectively, when $a_1 \neq a_2$ (Figure 6, bottom). We can see that when $a_1 = a_2$, most of the packets arrive at the receivers containing samples from two sources mixed together, while for $a_1 = 0.25$ (r_1), most of the packets contain samples from four sources, and for $a_2 = 0.75$ (r_2), more than one third of the arriving packets contain a mix of only two sources, and the rest only one source (no mixing). This shows that strategy 3 tries to stabilize at a target mixing level, which is characterized by the a parameter. The lowest the a parameter value is for a given session, the highest the mixing level which is achieved.

Finally, we vary the number of sessions in parallel using the third strategy, in three separate runs: during the first run all sessions have $a = 0.25$, during the second, $a = 0.5$, and the third, $a = 0.75$. The total average prices for memory and link buffer occupancy are depicted in Figure 7. We can see that the a parameter has a clear influence on the link prices, that increase with a for a given number of sessions, as expected. However it has little influence on the memory prices. This can probably be explained by the fact that, although the application avoids using the memory during a too long period due to delay constraints, at any time each session has at most one packet stored in memory. In the simulations shown, memory does not become a bottleneck, therefore the small impact on prices.

6. Conclusions and future work

We have presented a survey of current research on agent and active network techniques applied to adaptive applications, with special attention to optimization and market-based approaches. We have also described a model for trading resources inside an active network node, which draws many elements from agent technology. We have applied the model to a concast audio mixing application which trades off link resources against memory in the presence of bottleneck links. The

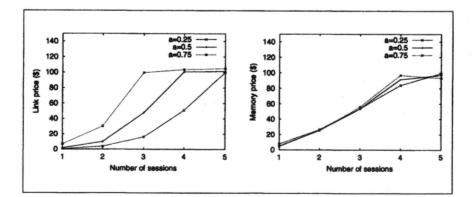

Figure 7. *Average link and memory prices when varying the number of sessions in parallel*

concast application is able to take congestion control decisions locally at each active node, such that no closed loop feedback between source and destination is needed. Using simulations, we studied three different strategies to make a decision on the amount of resources to use: two naive strategies based on the cheapest price, and a strategy that makes use of utility function weights. The results indicate that the first two strategies are already able to make improvements over the case when no congestion control is used, but they use resources inefficiently. The third strategy gives better results, achieving a stable and efficient sharing of resources.

We have several research directions to pursue: the most immediate one is to perform more complex simulations involving multiple node and link types, resource manager types, active and non-active nodes, different user strategies, etc. An implementation over a real active networking platform is also envisaged for the near future. We also plan to investigate the issues of dynamic resource manager upgrade with the help of mobile agents. The precise communication abstractions among the various kinds of agents need further attention too.

Acknowledgements

This work has been carried out within the TINTIN project funded by the Walloon region in the framework of the programme "Du numérique au multimédia".

REFERENCES

[ALE 99] D. S. Alexander, J. M. Smith, "The Architecture of ALIEN", *LNCS 1653, Proceedings of IWAN'99*, Berlin, Germany, June/July 1999, p. 1–12.

[ANA 00] K. G. Anagnostakis et al., "Scalable Resource Control in Active Networks", *LNCS 1942, Proceedings of IWAN 2000*, Tokyo, Japan, October 2000, p. 343–357.

[BOL 00] L. Bölöni, D. C. Marinescu, "Agent Surgery: The Case for Mutable Agents", *Proceedings of the Third Workshop on Bio-Inspired Solutions to Parallel Processing Problems (BioSP3)*, Cancun, Mexico, May 2000.

[BRE 00] J. Bredin et al., "A Game-Theoretic Formulation of Multi-Agent Resource Allocation", *Proceedings of the 2000 International Conference on Autonomous Agents*, Barcelona, Spain, June 2000.

[BRE 99] M. Breugst et al., "Grasshopper – An Agent Platform for Mobile Agent-Based Services in Fixed and Mobile Telecommunications Environments", In [HAY 99a], Chapter 14, p. 326–357.

[CAL 99] K. L. Calvert (ed.) et al., "Architectural Framework for Active Networks", (DARPA) AN Working Group, draft version 1.0, July 1999, work in progress.

[CAL 00] K. Calvert, "Toward an Active Internet", *Active and Programmable Networks Mini-conference, Networking 2000*, Paris, France, May 2000.

[CAM 99] A. T. Campbell, et al., "A Survey of Programmable Networks", *ACM SIGCOMM Computer Communication Review*, April 1999, p. 7–23.

[CHE 98] T. M. Chen, A. W. Jackson, "Active and Programmable Networks", Guest Editorial, *IEEE Network*, May/June 1998, p. 10–11.

[CLE 96] S. H. Clearwater (ed.), "Market-Based Control – A Paradigm for Distributed Resource Allocation", World Scientific Press, 1996.

[DEN 99] S. Denazis et al., "Designing Interfaces for Open Programmable Routers", *LNCS 1653, Proceedings of IWAN'99*, Berlin, Germany, June/July 1999, p. 13–24.

[FER 96] D. F. Ferguson, C. Nickolaou, J. Sairamesh, Y. Yemini, "Economic Models for Allocating Resources in Computer Systems", in [CLE 96], Chapter 7, p. 156–183.

[FLO 93] S. Floyd, V. Jacobsen, "Random Early Detection Gateways for Congestion Avoidance", *IEEE/ACM Transactions on Networking*, August 1993.

[FLO 00] S. Floyd, et al., "Equation-Based Congestion Control for Unicast Applications", *Proceedings of ACM SIGCOMM 2000*, Stockholm, Sweden, August 2000.

[GIB 99] M. A. Gibney, N. J. Vriend, J. M. Griffiths, "Market-Based Call Routing in Telecommunications Networks Using Adaptive Pricing and Real Bidding", *LNAI 1699, Proceedings of the IATA'99 Workshop*, Stockholm, Sweden, August 1999.

[GOP 00] R. Gopalakrishnan et al., "A Simple Loss Differentiation Approach to Layered Multicast", *Proceedings of IEEE INFOCOM 2000*, Tel-Aviv, Israel, March 2000.

[HAY 99a] A. L. G. Hayzelden, J. Bigham (eds), "Software Agents for Future Communication Systems", Springer-Verlag, 1999.

[HAY 99b] A. L. G. Hayzelden, et al., "Future Communication Networks using Software Agents", In [HAY 99a], Chapter 1, p. 1–57.

[HIC 99] M. Hicks et al., "PLANet: An Active Internetwork", *Proceedings of IEEE INFOCOM'99*, New York, 1999.

[HJA 00] G. Hjalmtysson, "The Pronto Platform: A Flexible Toolkit for Programming Networks using a Commodity Operating System", *Proceedings of IEEE OPENARCH 2000*, Tel-Aviv, Israel, March 2000, p. 98–107.

[JUN 00] K. Jun., L. Bölöni, D. Yau, D. C. Marinescu, "Intelligent QoS Support for an Adaptive Video Service", To appear in the *Proceedings of IRMA 2000*.

[KHA 00] I. El Khayat, G. Leduc, "Contrôle de Congestion pour la Transmission Multipoints en Couches", *JDIR'2000, 4ièmes Journées Doctorales Informatique et Réseaux*, Paris, France, November 2000.

[LOW 99] S. Low, D. E. Lapsley, "Optimization Flow Control, I: Basic Algorithm and Convergence", *IEEE/ACM Transactions on Networking*, 1999.

[MIZ 99] H. Mizuta, K. Steiglitz, E. Lirov, "Effects of Price Signal Choices on Market Stability", *4th Workshop on Economics with Heterogenous Interacting Agents*, Genoa, June 1999.

[NAJ 00] K. Najafi, A. Leon-Garcia, "A Novel Cost Model for Active Networks", *Proceedings of International Conference on Communication Technologies, World Computer Congress 2000*.

[NS 00] UCB/LBNL/VINT Network Simulator – ns (version 2),
http://www.mash.cs.berkeley.edu/ns/

[PAP 00] T. Papaioannou, "On the Structuring of Distributed Systems: The Argument for Mobility", PhD thesis, Loughborough University, February 2000.

[PET 00] L. Peterson (ed.) et al., "NodeOS Interface Specification", (DARPA) AN NodeOS Working Group, draft, January 2000, work in progress.

[PIN 98] R. S. Pindyck, D. L. Rubinfeld, "Microeconomics", 4th Edition, Prentice Hall International Inc., 1998.

[RUM 00] R. Stainov, J. Dumont, "Distributed Computations by Active Network Calls", *LNCS 1942, Proceedings of IWAN 2000*, Tokyo, Japan, October 2000, p. 45–56.

[SIV 00] R. Sivakumar, S. Han, V. Bharghavan, "A Scalable Architecture for Active Networks", *Proceedings of IEEE OPENARCH 2000*, Tel-Aviv, Israel, March 2000.

[SUG 99] K. Sugauchi et al., "Flexible Network Management Using Active Network Framework", *LNCS 1653, Proceedings of IWAN'99*, Berlin, Germany, June/July 1999, p. 241–248.

[TEN 97] D. L. Tennehouse et al., "A Survey of Active Network Research", *IEEE Communications Magazine*, Vol. 35, No. 1, p. 80–86, January 1997.

[TSC 93] C. Tschudin, "On the Structuring of Computer Communications", PhD thesis, University of Geneva, Switzerland, 1993.

[TSC 97] C. Tschudin, "Open Resource Allocation for Mobile Code", *Proceedings of the Mobile Agent '97 Workshop*, Berlin, Germany, April 1997.

[VAN 00] B. Vandalore et al., "A Survey of Application Layer Techniques for Adaptive Streaming of Multimedia", to appear in the *Journal of Real Time Systems*, 2000.

[WET 98] D. J. Wetherall, J. V. Guttag, D. L. Tennehouse, "ANTS: A Toolkit for Building and Dynamically Deploying Network Protocols", *Proceedings of IEEE OPENARCH'98*, San Francisco, USA, April 1998.

[YAM 96] H. Yamaki, M. P. Wellman, T. Ishida, "A Market-Based Approach to Allocating QoS for Multimedia Applications", *Proceedings of ICMAS'96*, Kyoto, Japan, December 1996.

[YAM 00a] L. Yamamoto, G. Leduc, "An Agent-Inspired Active Network Resource Trading Model Applied to Congestion Control", *LNCS 1931, Proceedings of MATA 2000*, Paris, France, September 2000.

[YAM 00b] L. Yamamoto, G. Leduc, "An Active Layered Multicast Adaptation Protocol", *LNCS 1942, Proceedings of IWAN 2000*, Tokyo, Japan, October 2000, p. 180–194.

Index

Innovative Technology Series
Information Systems and Networks

Other titles in this series

Advances in UMTS Technology

Edited by J. C. Bic and E. Bonek
£58.00   1903996147   216 pages   April 2002

The Universal Mobile Telecommunication System (UMTS), the third generation mobile system, is now coming into use in Japan and Europe. The main benefits – spectrum efficient radio interfaces offering high capacity, large bandwidths, ability to interconnect with IP-based networks, and flexibility of mixed services with variable data – offer exciting prospects for the deployment of these networks.

This publication, written by academic researchers, manufacturers and operators, addresses several issues emphasising future evolution to improve the performance of the 3rd generation wireless mobile on to the 4th generation. Outlining as it does key topics in this area of enormous innovation and commercial investment, this material is certain to excite considerable interest in academia and the communications industry.

The content of this book is derived from *Annals of Telecommunications*, published by GET, Direction Scientifique, 46 rue Barrault, F 75634 Paris Cedex 13, France.

Java and Databases

Edited by A. Chaudhri
£35.00   1903996155   136 pages   April 2002

Many modern data applications such as geographical information systems, search engines and computer aided design systems depend on having adequate storage management control. The tools required for this are called persistent storage managers. This book describes the use of the programming language Java in these and other applications.

This publication is based on material presented at a workshop entitled 'Java and Databases: Persistence Options' held in Denver, Colorado in November 1999. The contributions represent the experience acquired by academics, users and practitioners in managing persistent Java objects in their organisations.

For information about other engineering and science titles published by Hermes Penton Science, go to **www.hermespenton.com**

Quantitative Approaches in Object-oriented Software Engineering

Edited by F. Brito e Abreu, G. Poels, H. Sahraoui, H. Zuse
£35.00   1903996279   136 pages   April 2002

Software internal attributes have been extensively used to help software managers, customers and users characterise, assess and improve the quality of software products. Software measures have been adopted to increase understanding of how software internal attributes affect overall software quality, and estimation models based on software measures have been used successfully to perform risk analysis and to assess software maintainability, reusability and reliability. The object-oriented approach presents an advance in technology, providing more powerful design mechanisms and new technologies including OO frameworks, analysis/design patterns, architectures and components. All have been proposed to improve software engineering productivity and software quality.

The key topics in this publication cover metrics collection, quality assessment, metrics validation and process management. The contributors are from leading research establishments in Europe, South America and Canada.

Turbo Codes: Error-correcting Codes of Widening Application

Edited by M. Jézéquel and R. Pyndiah
£50.00   1903996260   206 pages   May 2002

The last ten years have seen the appearance of a new type of correction code – the *turbo code*. This represents a significant development in the field of error-correcting codes.

The decoding principle is to be found in an iterative exchange of information (*extrinsic information*) between elementary decoders. The turbo concept is now applied to block codes as well as other parts of a digital transmission system, such as detection, demodulation and equalisation.

Providing an excellent compromise between complexity and performance, turbo codes have now become a reference in the field, and their range of application is increasing rapidly to mobile communications, interactive television, as well as wireless networks and local radio loops. Future applications could include cable transmission, short distance communication or data storage.

This publication includes contributions from an internationally-based group of authors, from France, Sweden, Australia, USA, Italy, Germany and Norway.

The content of this book is derived from *Annals of Telecommunications*, published by GET, Direction Scientifique, 46 rue Barrault, F 75634 Paris Cedex 13, France.

For information about other engineering and science titles published by Hermes Penton Science, go to **www.hermespenton.com**

Millimeter Waves in Communication Systems

Edited by M. Ney
£50.00   1903996171   180 pages   May 2002

The topics covered in this publication provide a summary of major activities in the development of components, devices and systems in the millimetre-wave range. It shows that solutions have been found for technological processes and design tools needed in the creation of new components. Such developments come in the wake of the demands arising from frequency allocations in this range. The other numerous new applications include satellite communication and local area networks that are able to cope with the ever-increasing demand for faster systems in the telecommunications area.

The content of this book is derived from *Annals of Telecommunications*, published by GET, Direction Scientifique, 46 rue Barrault, F 75634 Paris Cedex 13, France.

Intelligent Agents for Telecommunication Environments

Edited by D. Gaïti and O. Martikainen
£35.00   1903996295   110 pages   June 2002

Telecommunication systems become more dynamic and complex with the introduction of new services, mobility and active networks. The use of artificial intelligence and intelligent agents, integrated reasoning, learning, co-operating and mobility capabilities to provide predictive control are among possible ways forward. There is a need to investigate performance, flow and congestion control, intelligent control environment, security service creation and deployment and mobility of users, terminals and services. New approaches include the introduction of intelligence in nodes and terminal equipment in order to manage and control the protocols, and the introduction of intelligence mobility in the global network. These tools aim to provide the quality of service and adapt the existing infrastructure to be able to handle the new functions and achieve the necessary co-operation between nodes. This book's contributors, who come from research establishments all over the world, address these problems and provide ways forward in this fast-developing area of intelligence in networks.

For information about other engineering and science titles published by Hermes Penton Science, go to **www.hermespenton.com**

Video Data

Edited by M-S Hacid and S. Hassas
£35.00   1903996228   128 pages   July 2002

With recent progress in computer technology and reduction in processing costs it is possible to store huge amounts of video data needed in today's communication applications. To obtain efficient use of such data efficient storage, querying and navigation of this data is needed. To meet the increasing demands of the new developments, new management techniques and tools need to be developed, and this publication addresses the application of the many research disciplines involved.

Multimedia Management

Edited by J. Neuman de Souza and N. Agoulmine
£40.00   1903996236   140 pages   July 2002

With the arrival of multimedia services via the network, the study of multimedia transmission over various network technologies has been the focus of interest for research teams all over the world.

The previously antagonistic QoS and IP-based network technologies are now part of an integrated approach, which are expected to lead to new intelligent approaches to traffic and congestion control, and to provide the end user with quality service customised multimedia communications. This publication emanates from papers presented at a Multimedia Management conference held in Paris in May 2000.

For information about other engineering and science titles published by Hermes Penton Science, go to **www.hermespenton.com**

Applications and Services in Wireless Networks

Edited by H. Afifi and D. Zeghlache
£58.00   1903996309   260 pages   July 2002

Emerging wireless technologies for both public and private use have led to the creation of new applications. These include the adaptation of current network management procedures and protocols and the introduction of unified open service architectures. Aspects such as accounting for multiple media access and QoS (Quality of Service) profiling must also be introduced to enable multimedia service offers, service management and service control over the wireless Internet. Security and content production are needed to foster the development of new services while adaptable applications for variable bandwidth and variable costs will open new possibilities for ubiquitous communications. In this book the contributors, drawn from a broad international field, address these prospects from the most recent perspectives.

Wireless Mobile Phone Access to the Internet

Edited by Thomas Noel
£40.00   1903996325   150 pages   August 2002

Wireless mobile phone access to the Internet will add a new dimension to the way we access information and communicate. This book is devoted to the presentation of recent research on the deployment of the network protocols and services for mobile hosts and wireless communication on the Internet.

A lot of wireless technologies have already appeared: IEEE 802.11b, Bluetooth, HiperLAN/2, GPRS, UTMS. All of them have the same goal: offering wireless connectivity with minimum service disruption between mobile handovers. The mobile world is divided into two parts: firstly, mobile nodes can be attached to several access points when mobiles move around; secondly, ad-hoc networks exist which do not use any infrastructure to communicate. With this model all nodes are mobiles and they cooperate to forward information between each other. This book presents these two methods of Internet access and presents research papers that propose extensions and optimisations to the existing protocols for mobility support.

One can assume that in the near future new mobiles will appear that will support multiple wireless interfaces. Therefore, the new version of the Internet Protocol (IPv6) will be one of the next challenges for the wireless community.

For information about other engineering and science titles published by Hermes Penton Science, go to **www.hermespenton.com**